# Engineering Patents

## Including Case Studies I & II

Robert Tata, B.M.S.E., P.E.

*AuthorHouse™*
*1663 Liberty Drive*
*Bloomington, IN 47403*
*www.authorhouse.com*
*Phone: 1-800-839-8640*



*Published by AuthorHouse 02/18/2014*

*ISBN: 978-1-4918-5571-3 (sc)*
*ISBN: 978-1-4918-5570-6 (e)*

*Library of Congress Control Number: 2014901246*



*Cover photos by Bobbie & Don Gfell's Sights and Sounds of Edison, Milan, Ohio edisonman1@aol.com*

*Interior design and layout by Ronald Pono.*

*This book is printed on acid-free paper.*

# INTRODUCTION

This book is written in two parts. The first part contains general information on patents and the first of two case studies pertaining to patents that were sought after by a major U.S. car company. The second part of this book describes the second case study from the same car company regarding a patent on rolling contact bearing competitive usage.

A patent is a legal document issued by the government that gives the engineering applicant exclusive rights to an invention for a period of 20 years. The invention must be something that is new and useful to society. An invention that is obvious to a person with average skill in a given field is not patentable. Patents are sometimes assigned to a company with one or more of its employees listed as the inventor. This is usually done as a prior condition of employment between the company and the employees. Patent laws give individuals and companies incentive to develop their innovative ideas for their own benefit and for the benefit of others. Some large corporations spend billions of dollars a year to develop useful and beneficial products that they otherwise may not have the incentive to do without the patent laws.

Patents can be traced as far back as 500 BC to Europe where the first patents were granted in Greece for "refinements in luxury". In the 1400's, in Italy, a patent was issued for a barge that carried marble. Patents were granted in England and France in the 1600's and 1700's. The first Congress of the United States adopted the Patent Act in 1790 and the first patent was issued on July 31, 1791 titled "in the making of Pot ash and Pearl ash by a new Apparatus and Process". It was signed by George Washington.

Treaties between the nations of the world including the U.S. and Canada have gone into effect whereby members of one nation may obtain patents in another nation in a manner similar to their own country's patent laws.

The word "patent" comes from the Latin word "patere" which means "to lay open". This refers to the fact that in order for a patent to be issued giving the inventor exclusive rights to his invention he must make his idea known to the general public. That is the reason that, in certain cases, some parties do not apply for patents for innovations which they consider too important to divulge for fear they will be infringed upon or ways will be found to circumvent them.

A patent gives the inventor exclusive rights to his invention; however, it does not give him the right to interfere with another patent. For example, many inventions are improvements of existing patented products. An inventor may not produce the item with his patented improvement without the consent of the original inventor assuming that the original patent is still in effect.

Patents may be sold, licensed, transferred, or even given away. It is common for corporations engaged in the same field of endeavor such as the automotive manufacturers to share patents for the mutual benefit of more than one company.

# Contents

## Part 1

# New Patent Law

A new patent law was signed by the President of the United States on September 16, 2011. One of the most significant parts of the legislation is that patents will now be awarded by the U.S. Patent Office from a "first to invent" system to a "first to file" system. This new procedure brings the U.S. in line with the rest of the world and, more importantly, ends the difficult and sometimes arduous task of trying to prove the first party to conceive of a contested patentable item. Another significant feature of the new law is that the Patent Office will now be allowed to set its own fees and manage its own budget while maintaining a position of being revenue neutral. This will help alleviate a growing backlog of business by hiring more patent examiners and procuring state-of-the-art technical equipment. The new law provides a discount for small companies and an even larger discount for individuals when filing patent applications. Also, another feature of the new law allows a patent to be challenged by going to the Patent Office rather than having to go to the courts.

# Patent Process

There are three types of patents - utility patents, design patents, and plant patents. Engineering patents usually fall into the category of utility patents. Utility patents involve materials, machines and components, and manufacturing parts and processes. Design patents involve only the appearance of an article. Plant patents, as the name suggests, are given to those that reproduce a new variety of plant.

When a new idea that is worth applying a patent for is conceived, a drawing or sketch should immediately be made and a description written that makes the idea understandable to people working in the same field. The documents should then be signed and dated by the inventor and two other people who understand the idea depicted by the sketch and the written description. If applicable, the drawing and written description should explain what is currently used, what the deficiencies are, and how they can be remedied by incorporating the new invention. The information should be presented in a detailed manner and include one or more claims made.

After all the preliminary work has been done, a Patent Attorney should be consulted. The Patent Attorney, for a fee, will make a search for "prior art" with the U.S. Patent Office. Prior art includes any patent that has been written (or is pending) that is the same as or similar to the one being applied for. Assuming that the prior art shows that the proposed patent is a new and novel idea, the Patent Attorney will prepare the patent application and submit it to the U.S. Patent Office who will examine the document and decide if a patent will be allowed. The latest fee to apply for a utility patent is $330 for an individual/s filing "on line" and $600 through the mail, $730 for a small company (under

500 employees) “on line” and $1000 through the mail, and $1320 for a large company “on line” and $2000 through the mail. There can also be search, examination, and issue fees. Patent Attorney fees can be much higher. If a patent is granted, there is a nominal multi-year maintenance fee. There are about 500,000 patents applied for and about 150,000 patents granted every year. There have been over 8 million patents issued by the U.S. Patent Office since its inception.

# CASE STUDY I

The following pages of this course examine a patent granted to a major U.S. car company. The patent was one of several ideas proposed in an attempt to improve the performance of automotive vehicle engine coolant pumps (waterpumps) which had become a burden on vehicle warranty costs to the company.

## Automotive Engine Coolant Pump:

Vehicle engine coolant pumps (waterpumps) circulate coolant through cored passages in the engine block where it absorbs heat. The fluid is then sent to the radiator where it is cooled and the cycle repeated. Waterpumps on the vehicles in question are mounted on the front of the engine block where they are driven by a belt from an engine driven pulley.

Figure 1 illustrates the general layout of a vehicle coolant pump. Drive pulley 1 has an integrated elongated shaft 2 which has pump rotor 3 mounted on the opposite end. The shaft is supported by double row ball bearing 4 which is mounted in waterpump housing 5 fastened to engine block 6. Fluid is pumped from the radiator through engine block opening 7 to engine block opening 8 where it flows through cooling passages and returns to the radiator where the cycle is repeated. The design necessitates the use of an all-important cartridge seal assembly located at 9 that prevents coolant from entering the bearing. The seal is the highest stressed and therefore one of the most critically analyzed components of the waterpump. Virtually all waterpump failures occur as a result of failure of this seal. The cartridge seal is a rather complex assembly that incorporates, among other things, two contacting circular elements, one stationary, and a second that rotates against the first (Cartridge seal assembly shown on Figure 8). The seal is designed so that the contacting surface between the two elements prevents coolant from entering the bearing. For the seal to operate satisfactorily, a small amount of coolant is allowed to enter the sealing zone for lubrication and cooling purposes. Occasionally, because of damage to the sealing surfaces from heat and the entrance of hard particle contaminants, an excessive amount of coolant vapors and liquid pass through the improperly functioning sealing elements to the next barrier which is the bearing seal located at 10. The bearing seal casing is designed to be mounted as a press fit in the bearing outer ring and has multiple lips that contact the rotating shaft. Excessive amounts of coolant vapor and liquid corrode the shaft, wear away the rubber sealing lips, and enter the bearing. There they act to impregnate the grease rendering it ineffective as a lubricant which, in turn, results in bearing failure. Replacing waterpumps can be a high warranty cost item to the auto companies.

# Figure 1

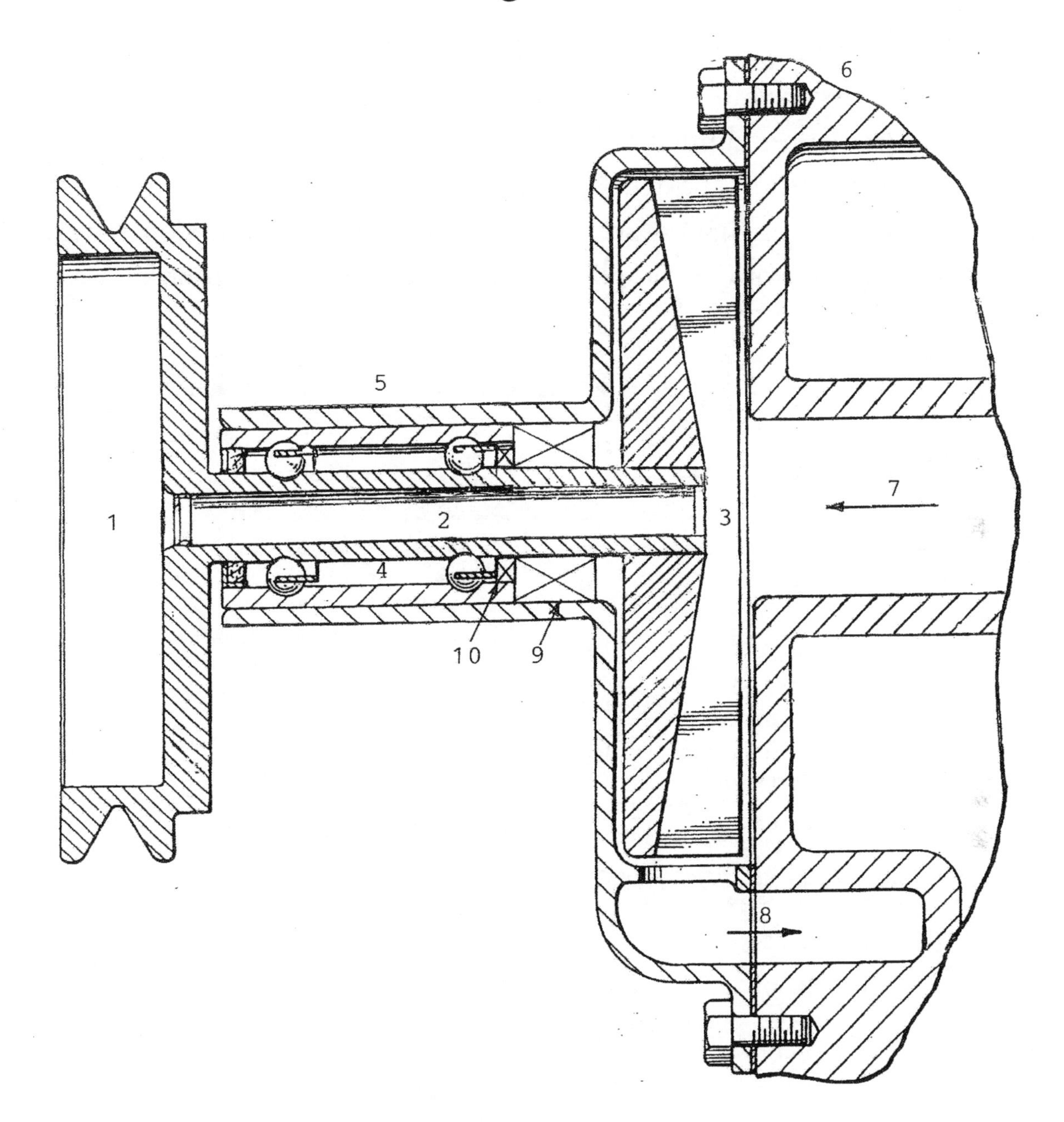

Magnetic Drive Vehicle Coolant Pump Patent:

Figure 2 is the first page of a U.S. Patent titled "Magnetic Drive Vehicle Coolant Pump". It was assigned to a U.S. car company with one of its employees listed as the inventor (names omitted). This first page contains the "Abstract" which, in one paragraph, defines the item being patented, its component parts, and the advantages that the design possesses. Below the abstract is a cross section drawing of the patented item with its components numbered for identification purposes.

## Figure 2

**United States Patent** [19]

[11] **Patent Number: 4,645,432**

[45] **Date of Patent: Feb. 24, 1987**

[54] MAGNETIC DRIVE VEHICLE COOLANT PUMP

[75] Inventor:

[73] Assignee:

[21] Appl. No.: 829,305

[22] Filed: Feb. 14, 1986

[51] Int. Cl.$^4$ .................... F04B 17/00; F04B 35/04

[52] U.S. Cl. .................................. 417/420; 415/10

[58] Field of Search ................ 417/420, 423 R, 362; 415/10, 122 R; 416/3; 310/104

[56] References Cited

U.S. PATENT DOCUMENTS

| | | | |
|---|---|---|---|
| 2,033,577 | 3/1936 | Hunter | 417/420 |
| 2,471,753 | 5/1949 | Johnston | 417/420 |
| 2,827,856 | 3/1958 | Zozulin | 417/420 |
| 2,939,974 | 6/1960 | Knight | 417/420 |
| 3,458,122 | 7/1969 | Andriussi | 416/3 |
| 3,627,445 | 12/1971 | Andriussi | 416/3 |
| 3,723,029 | 3/1973 | Laing | 417/420 |
| 3,732,445 | 5/1973 | Laing | 417/420 |
| 4,184,090 | 1/1980 | Taiani et al. | 417/420 |

*Primary Examiner*—Carlton R. Croyle
*Assistant Examiner*—Timothy S. Thorpe
*Attorney, Agent, or Firm*—Patrick M. Griffin

[57] ABSTRACT

A magnetic drive pump for use as a vehicle coolant pump. A fluid housing fixed to the engine block as an impeller mounted on the outside of a cylindrical support integrally stamped into a front wall of the housing. A pulley has a central hub that is rotatably mounted within the cylindrical support, coaxial with the impeller bearing. A web of the pulley and the impeller both face the housing front wall in closely spaced, parallel relation, with opposed matching magnetic drive elements. The structure is particularly simple and compact, and needs no cartridge or bearing seal.

**3 Claims, 1 Drawing Figure**

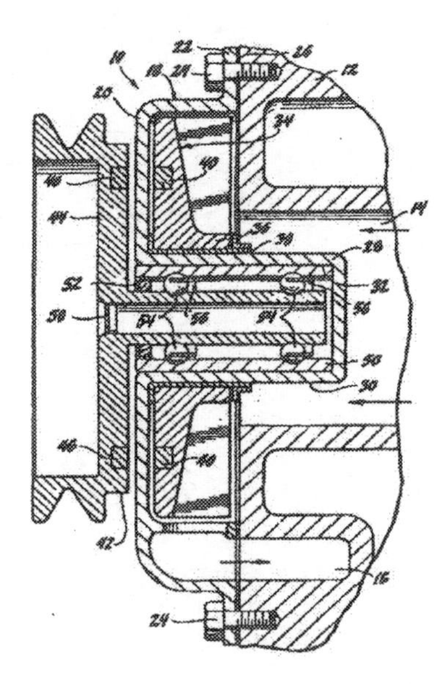

Figure 3 is the second page of the patent and is a larger scale replication of the drawing shown on Figure 2 that can better serve to explain the operation of the water pump. The main components of the waterpump are the pulley 42 which has a cylindrical protrusion that acts as the inner ring of a double row ball bearing 54. The double row ball bearing is mounted in a cylindrical section 28 of the pump housing 18. The pump housing is attached to the engine block 12 using a number of fasteners 24. The pulley is belt driven by an engine driven pulley (not shown). Embedded in the pulley are a number of magnets 46 whose attractive force acts on magnets 40 embedded

**Figure 3**

**U.S. Patent** **Feb. 24, 1987** **4,645,432**

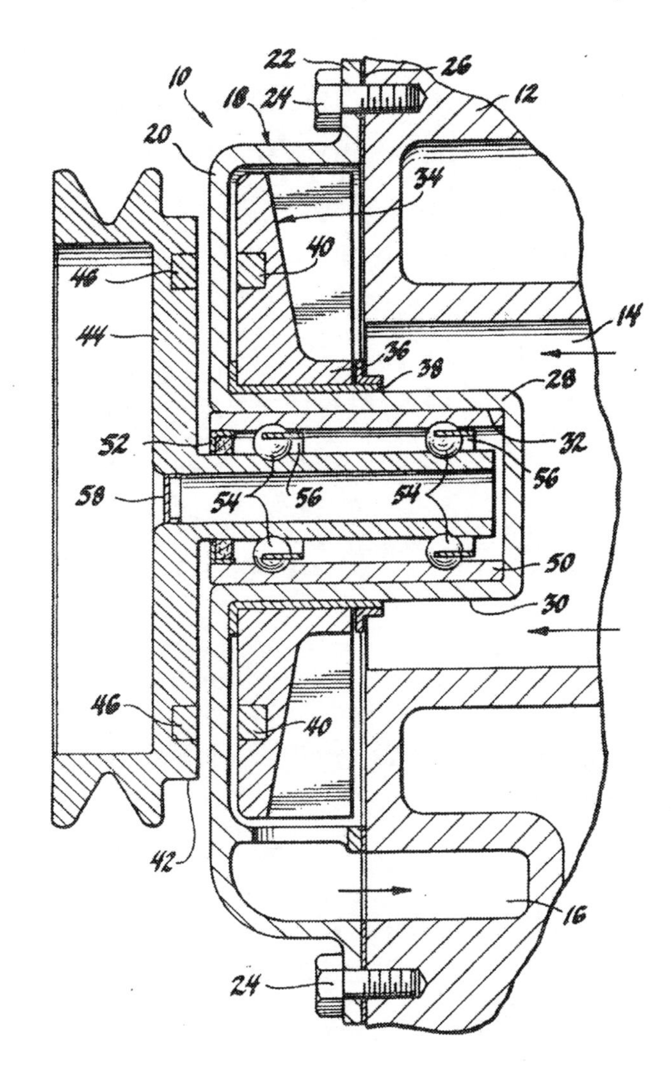

in the pump impeller 34 which serve to rotate the impeller. The rotating impeller pumps cooled fluid from the radiator through engine block passage 14 outward to engine block passage 16 where it serves to cool the engine. It is then sent back to the radiator and the cycle repeated. Because this invention uses magnets to rotate the impeller, there is no need for a cartridge seal; the design is simple; the design is axially compact.

Figure 4 contains the third page of the patent. As can be seen from all the text that is included on this page, (also on pages 8-10, Figures 5-7), Patent Attorneys (who write the actual document as a patent application), use elaborate phrasing and very descriptive language and are very careful in their selection of words in writing patents. This is done so that there is no question of what the patented item is, what it intends to accomplish, and how the accomplishments will be made - information that can be used as evidence in infringement cases.

Figure 4 contains the section titled "Background of the Invention" which explains, in lengthy discourse, the purpose and description of commonly used vehicle coolant pumps and problems associated with the cartridge seal. It then cites (4) U.S. Patents that are used to argue the case of why the subject patent should be granted:

# Figure 4



## MAGNETIC DRIVE VEHICLE COOLANT PUMP

This invention concerns vehicle coolant pumps in general, and specifically a coolant pump that uses a magnetic drive so as to provide a particularly simple and axially compact design.

## BACKGROUND OF THE INVENTION

Vehicle coolant pumps, often referred to as water pumps, are used to circulate coolant through the cooling passages of an engine block. They are generally operated by a driving member in the form of a pulley, which is in turn powered by a drive belt that runs off of the engine. It is necessary, of course, that the impeller of the pump be in communication with the coolant, in order to circulate it. The impeller is usually internal to a housing which is attached to the engine block and which encloses a space that communicates with the engine block cooling passages. A shaft or other member must be physically connected from the pulley to the impeller, which necessitates an opening physically through the housing. That opening must be sealed against the egress of coolant. The seal is highly stressed by the rapid rubbing rotation of the shaft that it surrounds, and by the heat of the coolant, coolant which may well contain abrasive particles, and will inevitably wear. The U.S. Pat. No. 3,632,220 to Lansinger et al illustrates well the problems with this conventional type of coolant pump. A generally cylindrical housing **9***a*, which stands out from the engine block **9**, has a shaft **11** supported by a bearing **12** passing through it. The shaft **11** is sealed with a complex seal assembly, generally referred to as a cartridge seal, made up of two seal members **26** and **27** spring loaded against one another. Although it is not numbered, one skilled in the art will recognize a weep hole through the housing *9a* opening to the ambient to vent the coolant that will invariably leak past the cartridge seal. In addition, a strong, and therefore highly frictional, bearing seal must be provided at the inner end of the bearing **12** to exclude leaking coolant from entering the bearing. Leaking coolant is the major cause of water pump bearing failure. It will also be noted that the pump disclosed is not particularly axially compact, as measured along the axis of the shaft **11**. The housing *9a* extends out from, not into, the engine block, and the bearing is spaced axially far away from the pump impeller **21**. The complexity of the cartridge seal, as well as the necessity of venting the leaking coolant, all militate against making the pump more axially compact by moving the cartridge seal and bearing back inside the block, where they would not be so accessible or easily vented.

It is known, in general, to operate a pump impeller located on one side of a closure with a driver located on the other side of the closure by the use of opposed magnetic elements on the pump impeller and driver. This avoids passing a shaft physically through the closure, and thus no seal is necessary around the shaft. Numerous patents exist in the field of magnetic drive pumps, all of which incorporate the basic feature just described, with the consequent advantage of avoiding a seal. They are directed to various narrow and specific structures, none of which one skilled in the art could apply, without the application of inventive effort, to use as a vehicle coolant pump. Most involve very different environments and problems, such as pumps to be used with a large tank of corrosive chemicals, where space is not a critical factor.

For example, the U.S. Pat. No. 4,304,532 to McCry shows such a pump with an impeller 38 operated by a driver 20 which that is in turn powered by a shaft 18 from a motor 12. There are no particular space limitations in such an environment, and the motor 12 can be axially far removed from the impeller 38 with no problem. Such is not the case in the cramped environment where a vehicle coolant pump is to be used. More importantly, a vehicle coolant pump cannot be powered directly by a separate power source like a motor, but must be run indirectly from the vehicle engine with a belt and pulley. That pulley must be rotatably supported and axially and radially located relative to the pump impeller. The motor 12 in McCry is large and stable, and has its own internal bearings, so it is a simple matter to rotatably support the shaft 18 and driver 20 relative to the impeller 38. Similarly usable structure is just not available in the environment of a vehicle coolant pump. Other patents illustrate the same point. The U.S. Pat. No. 3,802,804 to Zimmermann shows another magnetic tank pump, again with a large motor 40 to support and locate a driver 38 relative to an impeller 35, all occupying a relatively large space in an environment where space is not a limitation. Other patents in the same field, such as the U.S. Pat. No. 4,115,040 to Knorr, do not disclose anything about bearings to support the driver and impeller, taking it as a given that there would be more than sufficient space and structure in the particular environment to provide them.

## SUMMARY OF THE INVENTION

The subject invention provides a magnetic drive pump that is suitable for use as a vehicle coolant pump, thus eliminating the cartridge seal, and further provides such a pump that is particularly simple and axially compact.

The preferred embodiment of the coolant pump of the invention includes a fluid housing fixed to the engine block of a vehicle. The housing has a substantially planar front wall of non-magnetic material that encloses a space that is in communication with the cooling passages of the block. The front wall has an integral cylindrical support formed therein with its axis oriented substantially perpendicular to the front wall and extending into the interior of the fluid housing. The outer cylindrical surface of the cylindrical support, which is inside of the fluid housing and faces the coolant, is closed, and need not be sealed. The inner cylindrical surface opens out to the exterior of the fluid housing. A pump impeller inside the fluid housing has a central hub that coaxially surrounds the cylindrical support, and which is radially and axially supported on the outer surface thereof by by a flanged plain bearing. The impeller also has a magnetic portion that is thereby located closely facing and parallel to the inside of the front wall of the fluid housing.

A rotatable member, which, in the preferred embodiment is provided by a central hub that extends from the web of a driving pulley, is sized so as to fit coaxially within the cylindrical support of the fluid housing. The web of the pulley is substantially planar and generally perpendicular to its central hub, and includes a magnetic portion generally matching that of the impeller. In the preferred embodiment, the pulley hub actually fits within a cylindrical liner, which is in turn adapted to be press fitted within the inner surface of the fluid housing

# Figure 5



cylindrical support. Rolling bearing elements are disposed in the annular space between the pulley hub and the cylindrical liner to radially and axially support the pulley hub within the liner. Therefore, when the cylindrical liner with the rotatably supported pulley is press fitted within the cylindrical support, the planar web of the pulley is thereby located closely facing and parallel to the outside of the front wall of the fluid housing. The magnetic portions of the pulley web and impeller are thereby located in opposition to each other across the front wall. The pulley is thus able to magnetically drive the impeller when the pulley is rotated by the vehicle engine through a drive belt. A very simple structure is thus provided with no necessity of a cartridge seal, or for a seal to exclude leaking coolant from the bearing elements, or for weep holes to the ambient to vent leaking coolant. Furthermore, the particular spatial arrangement, with the pulley and impeller bearings located one within the other, and with the pulley web and impeller in closely facing opposition across the fluid housing front wall, gives a particularly axially compact unit. Several advantages, therefore, are cooperatively provided by the same structure.

It is, therefore, a broad object of the invention to provide a vehicle coolant pump that is magnetically driven, thus eliminating the cartridge seal, and to do so with a structure that is well suited to that specific environment, being particularly simple and axially compact.

It is another object of the invention to provide such a vehicle coolant pump structure in which a fixed fluid housing has a substantially planar wall of non-magnetic material with a cylindrical support extending from the fluid housing wall into the interior of the fluid housing and axially oriented substantially perpendicular to the housing wall, with a closed outer cylindrical surface inside of the fluid housing and an inner cylindrical surface opening to the exterior of the fluid housing, and in which a pump impeller inside the fluid housing coaxially surrounds and is radially and axially supported by the outer cylindrical surface of the cylindrical support, with the impeller having a magnetic portion that is thereby located closely facing and parallel to the inside of the fluid housing wall, and in which a rotatable member sized so as to fit coaxially within the cylindrical support and radially and axially supported by the cylindrical support inner surface has a driving member attached thereto, a driving member that has a substantially planar web located closely facing and parallel to the outside of the fluid housing wall with a magnetic portion of the web in opposition to the impeller magnetic portion so as to drive the impeller when the driving member rotates, the driving member, housing wall and pump impeller occupying a compact axial space by virtue of their relative location.

It is yet another object of the invention to provide such a vehicle coolant pump structure in which a rotatable driving member has a central cylindrical hub sized so as to fit coaxially within the fluid housing cylindrical support and radially and axially supported by the cylindrical support inner surface, so that a web of the driving member is thereby located closely facing and parallel to the outside of the fluid housing wall with its magnetic portion in opposition to the impeller magnetic portion so as to drive the impeller when the driving member rotates.

It is still another object of the invention to provide such a vehicle coolant pump structure in which the hub of the driving member fits within, and is radially and axially supported by bearing elements within, a cylindrical liner which is in turn adapted to be press fitted within the inner surface of the cylindrical support, whereby the cylindrical liner, with the rotatably supported driving member, may be press fitted within the fluid housing cylindrical support, thereby locating the planar web closely facing and parallel to the outside of the fluid housing wall with the magnetic portions of the impeller and pulley web in opposition.

## DESCRIPTION OF THE PREFERRED EMBODIMENT

These and other objects and features of the invention will appear from the following written description and the drawing, which shows a cross section of the preferred embodiment in place on a portion of a vehicle engine block.

Referring to the drawing, the preferred embodiment of the subject invention, designated generally at **10**, provides a magnetic drive pump that is suitable for use as a vehicle coolant pump, thus eliminating the main seal, and further provides such a pump that is particularly simple and axially compact. The coolant pump of the invention **10**, is shown attached to a portion of a vehicle engine block, designated generally at **12**. Engine block **12**, as is typical, is cast with cooling passages, an inlet passage designated at **14** and an outlet passage designated at **16**. Coolant flows through the passages **14** and **16**, pumped by the coolant pump **10**, as indicated by the arrows. The coolant pump **10** includes a fluid housing, designated generally at **18**. Fluid housing **18** is stamped of aluminum or other suitable non-magnetic material, and includes a generally planar front wall **20** and a peripheral flange **22**. When it is fixed with bolts **24** and a gasket **26** to block **12**, fluid housing **18** encloses a space that is in communication with the cooling passages **14** and **16**. That fixing does not occur until after other assembly steps described below have been completed, however. A cylindrical support, designated generally at **28**, is integrally stamped into front wall **20** and extends inwardly therefrom with its axis generally perpendicular thereto. The outer cylindrical surface **30** of the cylindrical support **28**, which is inside of the fluid housing **18** and faces the coolant, is closed, and need not be sealed. The inner cylindrical surface **32** opens out to the exterior of the fluid housing **18**. A pump impeller, designated generally at **34**, is located inside the fluid housing. Impeller **34** has a central hub **36** that coaxially surrounds the cylindrical support **28**, and which is radially and axially supported on the outer cylindrical surface **30** thereof by a flanged plain bearing **38**. The impeller **34** has a magnetic portion **40** that is thereby located closely facing and parallel to the inside of the front wall **20** of the fluid housing **18**. Impeller **34** would not be added until after a prior step described below, however.

A driving member is provided by a pulley, designated generally at **42**, which would be powered by a belt driven by the vehicle engine, not shown. Pulley **42** could be formed of 1070 steel or other suitable material, and includes a generally planar web **44** into which is set a magnetic portion **46** that generally matches the magnetic portion **40** of impeller **34**. A rotatable member is provided by a central hub **48** that extends from the web **44**, generally perpendicular thereto. Hub **48** is sized so as to fit coaxially within the fluid housing cylindrical support **28**. In the preferred embodiment, the pulley hub **48** actually fits within a separate cylindrical liner **50** of bearing quality steel, which is in turn sized so that it can

# Figure 6



be press fitted within the inner cylindrical surface **32** of the cylindrical support **28**, with an annular space therebetween.

The manufacturing and assembly process of the coolant pump **10** is as follows. Ball pathways are formed in the outer and inner surfaces respectively of hub **48** and liner **50**, and induction hardened by conventional means. A dust seal **52** is pressed into one end of liner **50**. Then, two rows of bearing balls **54** are placed in through the unobstructed right end of the annular space between liner **50** and hub **48**. The balls **54** are conrad assembled between the pathways, and standard snap-in separators **56** added. This serves to radially and axially support the pulley hub **48** within the liner **50**, and creates a separately handled subunit made up of the liner **50** and the pulley **42** rotatably supported thereto. Then, by heat expanding the cylindrical support **28**, liner **50** may be press fitted easily thereinto. When the cylindrical liner **50** has been so assembled, the pulley web **44** is thereby located closely facing and parallel to the outside of the front wall **20** of the fluid housing **18**. The plain bearing **38** and impeller **34** may then be added, and the impeller magnetic portion **40** will thereby be located in opposition to the pulley magnetic portion **46**, facing it across the non-magnetic front wall **20**. Finally, the fluid housing **18** is bolted in place as described above. A dust plug **58** may be added to the center hole of hub **48**, if desired.

Once the above assembly steps are completed, it will be understood that pulley **42** will able to magnetically drive the impeller **34** when it is rotatably driven by the vehicle engine. The driven impeller **34** will circulate the coolant in the pattern shown by the arrows. The use of this indirect, magnetic drive makes several things possible. It allows for a very simple structure, compared to conventional, directly driven vehicle coolant pumps. No cartridge seal or tight bearing seal is necessary, giving a very low friction and low torque structure with almost no parts susceptible to wear or failure. Nor are weep holes opening to the ambient out of the housing necessary. Eliminating these conventional items allows the pulley bearings **54** to be moved axially inboard, inside of and occupying essentially the same axial space as the impeller bearing **38**. This gives a highly axially compact unit, which is very advantageous in the cramped environment of increasingly smaller cars. Alternatively, the particular compact spatial arrangement may be thought of as serving to bring the pulley web **44** and impeller **34** into sufficiently closely facing relation to allow the matching magnetic portions **46** and **40** to operate. However the invention is conceptualized, it is apparent that a number of advantages cooperatively flow from a very simple and tightly interacting structure.

Variations of the preferred embodiment disclosed may be made within the sprit of the invention. For example, a separate shaft could replace the hub **48**, with a pulley attached separately to it, although that would mean more total parts. While the integral ball pathways on the hub **48** are practical, a separable raceway could be used instead, if desired. Or, it is possible that an integral ball pathway could be formed on the inner surface of support **28**, as well as on the hub **48**, especially if the pulley **42** were made separable from its hub **48**. This would allow conrad assembly of the balls directly into the cylindrical support **28** from the left end of the annular space. This would eliminate the liner **50**, but the liner **50** is desirable since support **28** is unlikely to be formed of bearing quality material. It is also advantageous to have the easily handled subunit comprised of pulley **42** and liner **50**, as described. Bearing elements other than balls **54** could be used, as well, although balls are particularly easy to assemble in the environment disclosed. Therefore, it will be understood that it is not intended to limit the scope of the invention to just the preferred embodiment disclosed.

The embodiments of the invention in which an exclusive property or privilege is claimed are defined as follows:

**1.** An axially compact magnetic drive pump for use as a vehicle coolant pump, comprising,

a fixed fluid housing having a substantially planar wall of non-magnetic material with a cylindrical support extending from said fluid housing wall into the interior of said fluid housing, said cylindrical support having its axis oriented substantially perpendicular to said housing wall and having a closed outer cylindrical surface inside of said fluid housing and an inner cylindrical surface opening to the exterior of said fluid housing,

a pump impeller inside said fluid housing coaxially surrounding and radially and axially supported by the outer cylindrical surface of said cylindrical support, said impeller having a magnetic portion that is thereby located closely facing and parallel to the inside of said fluid housing wall,

a rotatable member sized so as to fit coaxially within said fluid housing cylindrical support and radially and axially supported by said cylindrical support inner surface, and,

a driving member having a substantially planar web with a magnetic portion, said driving member being attached to said rotatable member so that said planar web is located closely facing and parallel to the outside of said fluid housing wall with its magnetic portion in opposition to said impeller magnetic portion so as to drive said impeller when said driving member rotates, said driving member, housing wall and pump impeller occupying a compact axial space by virtue of their relative location.

**2.** An axially compact magnetic drive pump for use as a vehicle coolant pump, comprising,

a fixed fluid housing having a substantially planar wall of non-magnetic material with a cylindrical support extending from said fluid housing wall into the interior of said fluid housing, said cylindrical support having its axis oriented substantially perpendicular to said housing wall and having a closed outer cylindrical surface inside of said fluid housing and an inner cylindrical surface opening to the exterior of said fluid housing,

a pump impeller inside said fluid housing coaxially surrounding and radially and axially supported by the outer cylindrical surface of said cylindrical support, said impeller having a magnetic portion that is thereby located closely facing and parallel to the inside of said fluid housing wall, and,

a rotatable driving member having a central cylindrical hub sized so as to fit coaxially within said fluid housing cylindrical support and radially and axially supported by said cylindrical support inner surface, said driving member also having a substantially planar web with a magnetic portion that is thereby located closely facing and parallel to the outside of said fluid housing wall with its magnetic portion in opposition to said impeller magnetic portion so as

# Figure 7





to drive said impeller when said driving member rotates, said driving member, housing wall and pump impeller occupying a compact axial space by virtue of their relative location.

3. An axially compact magnetic drive pump for use as a vehicle coolant pump, comprising,

a fixed fluid housing having a substantially planar wall of non-magnetic material with a cylindrical support extending from said fluid housing wall into the interior of said fluid housing, said cylindrical support having its axis oriented substantially perpendicular to said housing wall and having a closed outer cylindrical surface inside of said fluid housing and an inner cylindrical surface opening to the exterior of said fluid housing,

a pump impeller inside said fluid housing coaxially surrounding and radially and axially supported by the outer cylindrical surface of said cylindrical support, said impeller having a magnetic portion that is thereby located closely facing and parallel to the inside of said fluid housing wall,



a cylindrical liner adapted to be press fitted within the inner surface of said cylindrical support,

a driving member having a central cylindrical hub sized so as to fit coaxially within said cylindrical liner and annularly spaced therefrom, said driving member also having a substantially planar web with a magnetic portion, and,

rolling bearing elements disposed in said annular space to radially and axially support said driving member hub coaxially within said cylindrical liner, whereby said cylindrical liner with said driving member rotatably supported therein may be press fitted within said fluid housing cylindrical support, thereby locating said planar web closely facing and parallel to the outside of said fluid housing wall with its magnetic portion in opposition to said impeller magnetic portion so as to drive said impeller when said driving member rotates, said driving member, housing wall and pump impeller occupying a compact axial space by virtue of their relative location.

* * * * *

5 10 15 20 25 30 35 40 45 50 55 60 65

1) U.S. Patent 3,632,220, "Coolant Pump": The first page of this patent is shown as Figure 8. The abstract describes the function and operation of the Invention" which explains, in lengthy discourse, the purpose and description of commonly used vehicle coolant pumps and problems associated with the cartridge seal. It then cites (4) U.S. Patents that are used to argue the case of why the subject patent conventional engine coolant pumps (with drawing) and the features of the invention which serve to cool the cartridge seal and prevent particles such as casting core sand from causing damage to it. This patent is cited to illustrate the vulnerability of the cartridge seal in the design of conventional engine coolant pumps which adds creditability to the design of the subject patent which eliminates the cartridge seal altogether. It is also cited to show how conventional coolant pumps with direct drive from the pulley through the bearing shaft to the pump impeller, and possessing a cartridge seal, are inherently more complex and axially longer than the subject patent.

The drawing on Figure 8 shows the complexity of the cartridge seal. Part number 26 is the rotating sealing member and 27 is the stationary sealing

## Figure 8

**United States Patent** [11] **3,632,220**

| | | |
|---|---|---|
| [72] | Inventors | **Jere R. Lansinger Westland; James E. MacAfee, Troy, both of Mich.** |
| [21] | Appl. No. | **67,436** |
| [22] | Filed | **Aug. 27, 1970** |
| [45] | Patented | **Jan. 4, 1972** |
| [73] | Assignee | **Chrysler Corporation Highland Park, Mich.** |

[54] **COOLANT PUMP**
**9 Claims, 4 Drawing Figs.**

[52] **U.S. Cl.** .......... **415/112,** 415/170

[51] **Int. Cl.** .......... **F01d 11/00**

[50] **Field of Search** .......... 415/110, 111, 112, 170

[56] **References Cited**

UNITED STATES PATENTS

| | | | |
|---|---|---|---|
| 2,203,525 | 6/1940 | Dupree, Jr. | 415/111 |
| 2,352,636 | 7/1944 | Jackman | 415/111 |
| 3,074,690 | 1/1963 | Henny | 415/111 |
| 2,769,390 | 11/1956 | Heimbach | 415/111 |

*Primary Examiner*—C. J. Husar
*Attorney*—Talburtt and Baldwin

**ABSTRACT:** A rotary coolant pump for an internal-combustion engine has a coaxial annular centrifuge cavity and a scoop operable upon rotation of the pump to collect fluid coolant into the cavity under pressure through an inlet port located between radially outer and inner discharge ports of the cavity. The inner discharge port communicates with an annular seal for the pump journal and is isolated from high-velocity coolant flow within the cavity by a region therein of enlarged cross-sectional area which reduces the coolant velocity at a location spacing the inner discharge port from the inlet and outer discharge ports. Coolant and high-density particles such as core sand are thus discharged radially outwardly from the cavity by centrifugal force through the outer discharge port, whereas clean pressurized coolant flows through the inner discharge port to cool the seal.

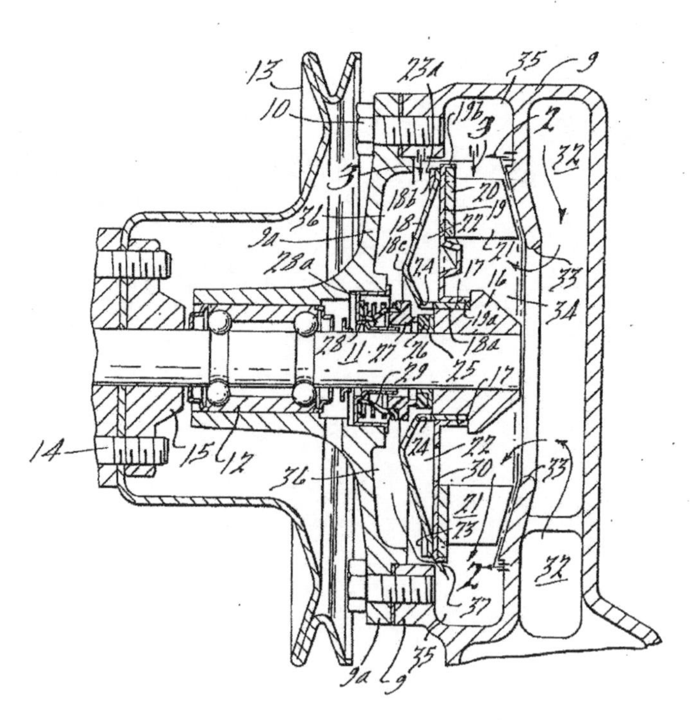

member. Number 28 is a spring which acts on the stationary ring to apply sealing pressure to the sealing elements. Also shown is annular boot 29 which prevents leakage around the stationary side of the seal and spring retainer 28a which serves to mount the cartridge assembly inside the pump housing bore. A simplified representation of the bearing seal (not numbered) is shown just to the left of the cartridge seal.

2) U.S. Patent 3,802,804, "Magnetically Coupled Pump Structure": The first page of this patent is shown as Figure 9. It describes a magnetically driven pump mounted on the side of a large tank that contains very hot liquids and corrosive chemicals. The reason for citing this patent is to acknowledge that the magnetically driven feature of the subject patent is itself not new but the fact that it is of simple design and axially compact (both extremely important advantages when mounting to the front of a vehicle engine block) separate it from this cited patent which has virtually no space limitations.

3) U.S. Patent 4,304,532, "Pump Having Magnetic Drive":

# Figure 9

**United States Patent** [19] [11] **3,802,804**

**Zimmermann** [45] **Apr. 9, 1974**

[54] **MAGNETICALLY COUPLED PUMP STRUCTURE**

[75] Inventor: **Frederick N. Zimmermann,** Deerfield, Ill.

[73] Assignee: **March Manufacturing Company,** Skokie, Ill.

[22] Filed: **July 21, 1967**

[21] Appl. No.: **655,109**

[52] **U.S. Cl.** .......... **417/360,** 417/420

[51] **Int. Cl.** .......... **F04b 35/04**

[58] **Field of Search**....... 103/87, 87 M; 230/15 MC; 310/104; 64/28 M; 192/84 PM; 417/420, 360

[56] **References Cited**

UNITED STATES PATENTS

| | | | |
|---|---|---|---|
| 2,970,548 | 2/1961 | Berner | 417/420 |
| 3,398,695 | 8/1968 | Pritz | 103/87 M |
| 3,411,450 | 11/1968 | Clifton | 103/87 M |
| R26,094 | 10/1966 | Zimmermann | 103/87 M |
| 2,855,141 | 10/1958 | VanRijn | 230/117 A |
| 3,144,573 | 8/1964 | Bergey et al. | 310/104 |
| 3,299,819 | 1/1967 | McCoy | 103/87 M |

*Primary Examiner*—William L. Freeh
*Attorney, Agent, or Firm*—Callard Livingston

[57] **ABSTRACT**

Centrifugal pump apparatus of the magnetically-coupled type suitable for tank mounting and also including improvements in impeller and spindle structures of general application and for use with very hot liquids and corrosive chemicals. A cup-shaped motor mounting bell is provided for insertion, open side out, into a hole in the side of a tank and has a flange attaching to the tank wall. The motor is supported on mounting formations inside of the large well afforded by the bell. A smaller cup-shaped magnet well is formed by recessing inwardly on the bottom of the cup-shaped bell, which is also provided with sealing land closing and sealing with the open side of the pump housing. The driven coupling magnet of the pump fits into the small magnet well which is surrounded by the larger motor-driven magnet in the larger well. The pump impeller and magnet rotate on spindle means supported at both ends and which may be integrally conformed at one end with a part of the pump structure.

**3 Claims, 9 Drawing Figures**

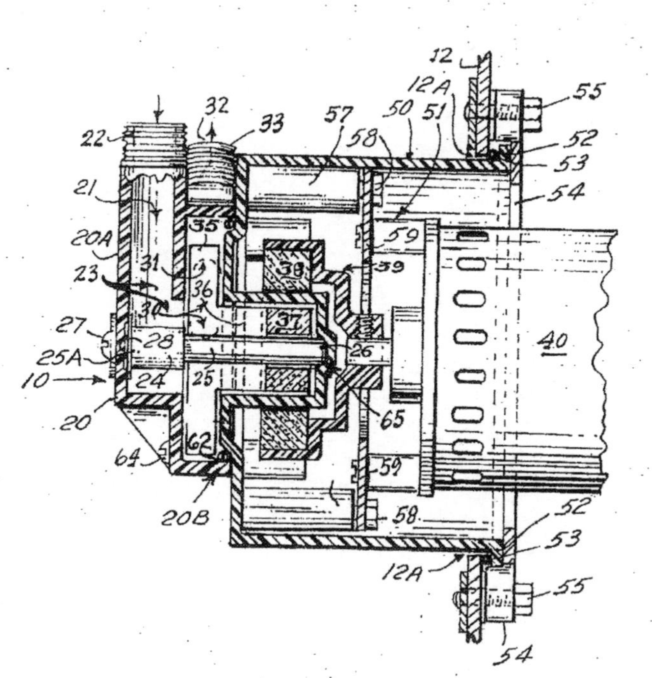

The first page of this patent is shown on Figure 10. It is cited for the same reason that the above patent is cited. It has a magnetic drive but there are no space limitations. Also, it is powered by an electric motor while a vehicle coolant pump is powered by an engine driven belt.

4) U.S. Patent 4,115,040, "Permanent Magnet Type Pump": The first page of this patent is shown on Figure 11. It is cited again for the same reason that above patents are cited. It has a magnetic drive but there are no space limitations.

The next section of the subject patent text begins on Figure 4 (page 7) and is titled "Summary of the Invention". It is an extremely elaborately worded section which explains, in detail, the object of every facet of every part of the invention. Again, it is included in the patent so that there will be no mistake of what the invention is comprised of and what its patentable features are.

The next section of the patent text is contained on Figures 5 to 7 (pages 8-10) and is titled "Description of the Preferred Embodiment".

# Figure 10

**United States Patent** [19]
McCoy

[11] **4,304,532**
[45] **Dec. 8, 1981**

[54] **PUMP HAVING MAGNETIC DRIVE**

[76] Inventor: **Lee A. McCoy**, 2605 Garfield St., San Mateo, Calif. 94403

[21] Appl. No.: **104,545**

[22] Filed: **Dec. 17, 1979**

[51] **Int. Cl.**[3] .......................................... **F04D 13/02**
[52] **U.S. Cl.** .................................. **417/420**; 310/104
[58] **Field of Search** ..................... 417/420; 64/28 M; 192/84 PM; 310/104

[56] **References Cited**

U.S. PATENT DOCUMENTS

| | | | |
|---|---|---|---|
| 2,669,668 | 2/1954 | Okulitch et al. | 417/420 X |
| 3,299,819 | 1/1967 | McCoy | 417/420 |
| 3,420,184 | 1/1969 | Englesberg | 417/420 |
| 3,512,901 | 5/1970 | Law | 417/420 X |

*Primary Examiner*—Richard E. Gluck
*Attorney, Agent, or Firm*—T. R. Zegree

[57] **ABSTRACT**


A fluid-handling apparatus having a magnetic drive comprises a rotatable driver member and a driven member provided with blades for imparting motion to a fluid and mounted on a stationary shaft forming a one-piece unit with a thin diaphragm positioned between said two members which comprise a plurality of permanent magnets. Each magnet has a central aperture dimensioned so that the attracting force is substantially equal throughout the body of the magnet.


**9 Claims, 5 Drawing Figures**

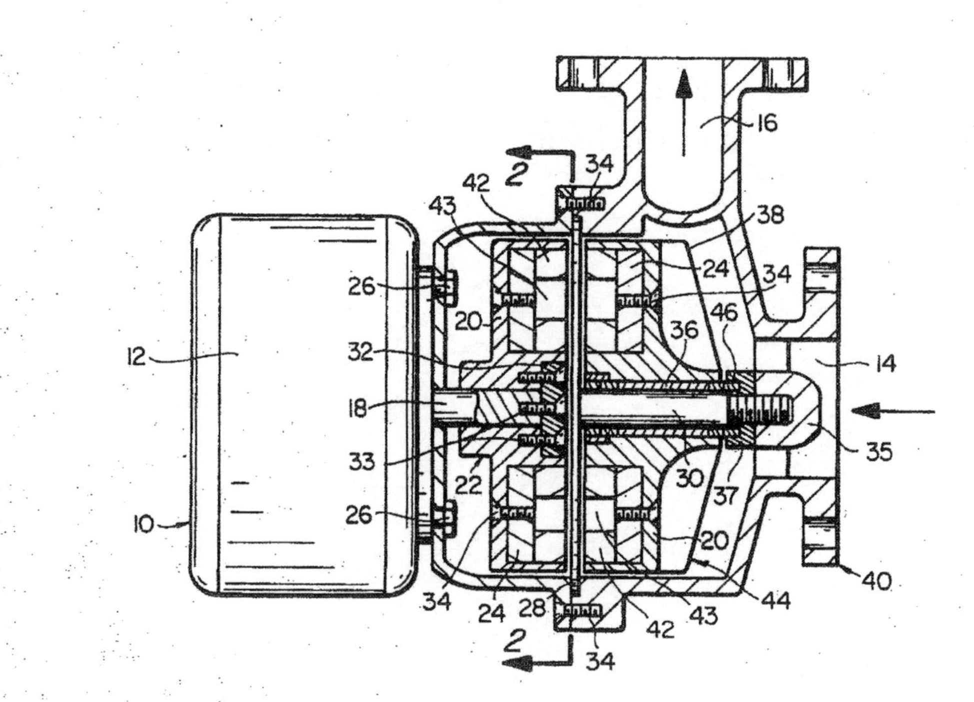


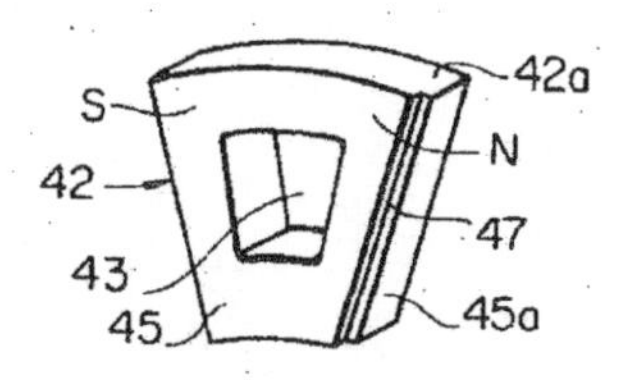

# Figure 11

# United States Patent [19]

**Knorr**

[11] **4,115,040**

[45] **Sep. 19, 1978**

[54] **PERMANENT MAGNET TYPE PUMP**

[75] Inventor: **Manfred Knorr,** Bochum-Wattenscheid, Germany

[73] Assignee: **Franz Klaus-Union,** Germany

[21] Appl. No.: **800,219**

[22] Filed: **May 25, 1977**

[30] **Foreign Application Priority Data**

May 28, 1976 [DE] Fed. Rep. of Germany ....... 2624058

[51] **Int. Cl.**² .............................................. **F04B 17/00**

[52] **U.S. Cl.** .................................. **417/420;** 64/28 M; 192/84 PM; 310/104

[58] **Field of Search** ....................... 417/420; 64/28 M; 192/84 PM; 310/103, 104, 105

[56] **References Cited**

U.S. PATENT DOCUMENTS

| | | | |
|---|---|---|---|
| 3,299,819 | 1/1967 | McCoy | 417/420 |
| 3,378,710 | 4/1968 | Martin, Jr. | 417/420 |
| 3,411,450 | 11/1968 | Clifton | 417/420 |
| 3,647,314 | 3/1972 | Laessig | 417/420 |
| 4,013,384 | 3/1977 | Oikawa | 417/420 |
| 4,065,234 | 12/1977 | Yoshiyuki et al. | 417/420 |

FOREIGN PATENT DOCUMENTS

2,534,740 2/1977 Fed. Rep. of Germany .......... 417/420

*Primary Examiner*—C. J. Husar
*Attorney, Agent, or Firm*—Woodard, Weikart, Emhardt & Naughton

[57] **ABSTRACT**

Disclosed is a permanent magnet pump in which the pump impeller and the interior rotor of a permanent magnet driving means receives drive torque transmitted in synchronism by an exterior rotor. The exterior rotor is positioned, axially in one form and radially in another form, opposite the interior rotor with an air gap defined between them. Thin, plate-like permanent magnets carried by the rotors face each other across the air gap. The pump impeller shaft and the interior rotor are housed and supported in a common space which is sealed from the exterior by a partition of non-magnetizable material extending through the air gap.

**8 Claims, 4 Drawing Figures**

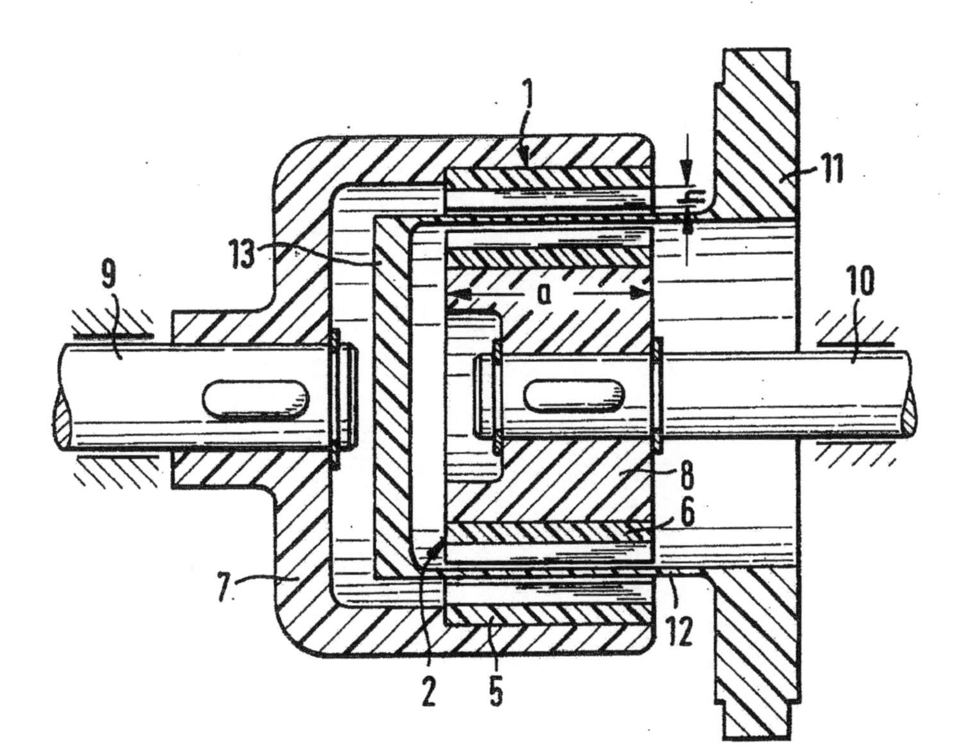

It explains in great detail the function of every part of the patent using the part numbers that are given on the drawing on the first two pages of the invention (Figures 2&3). It ends proclaiming each of the three claims made by this invention which are enumerated as follows:

Claim 1) This claim describes the invention as comprising the pump housing and the pump impeller; however, the pulley and the hub are described as separate pieces which is not the way they are shown on the drawing but is a credible option.

Claim 2) This claim is the same as claim 1 except the pulley and the hub are listed as one piece which is the way they are shown on the drawing.

Claim 3) This claim is the same as claim 2 except the double row "rolling" bearing is mentioned for the first time which is the way the invention is shown on the drawing. The pump could conceivable be made without a rolling bearing where arguments could be made that such a design variation is intended by claims 1 and 2 where no mention is made of the rolling bearing. Notice that the word "rolling" was used to describe the bearing which is shown as a ball bearing on the drawing but because it was referred to as a "rolling" bearing indicates that the use of a roller bearing is also part of the invention.

# QUIZ

1) A patent must be something:
    a) New
    b) Useful
    c) Not obvious
    d) All of the above

2) Patents can be traced as far back as 500 BC in:
    a) France
    b) England
    c) Italy
    d) Greece

3) Under the new patent law cases can be held in:
    a) Criminal courts
    b) Federal courts
    c) The U.S. Patent Office
    d) None of the above

4) Engineering patents are classified as:
    a) Utility patents
    b) Design patents
    c) Plant patents
    d) None of the above

5) When a patentable idea is thought of:
    a) A sketch should be made
    b) A description should be written
    c) The document should be signed and dated
    d) All of the above

6) Prior art is:
    a) Patents granted to a famous inventor
    b) Patents similar to the one being applied for
    c) Drawings being made for the patent sought
    d) None of the above

7) Vehicle Coolant Pumps most importantly assist:
    a) To cool the vehicle passengers
    b) To cool the transmission
    c) To cool the engine
    d) All of the above

8) Vehicle coolant pumps incorporate a:
    a) Double row bearing
    b) Drive pulley
    c) Pump impeller
    d) All of the above

9) The cartridge seal seals:
    a) Engine lubricant
    b) Engine fuel
    c) Engine coolant
    d) None of the above

10) The cartridge sealing elements consist of:
   a) Two face contacting discs
   b) A Single lip seals
   c) Multiple lip seals
   d) A labyrinth seal

11) The double row ball bearing seal consists of:
   a) Two face contacting discs
   b) A single lip seal
   c) A multiple lip seal
   d) A labyrinth seal

12) The "Abstract" of a patent:
   a) Defines the item being patented
   b) Names the component parts of the invention
   c) Identifies the advantages that the patent possesses
   d) All of the above

13) The patented waterpump:
   a) Uses magnets to rotate the impeller
   b) Has no need for a cartridge seal
   c) Is axially compact
   d) All of the above

14) The magnets in the patent are used to:
   a) Attract foreign debris from the coolant
   b) Drive the coolant pump
   c) Mount the coolant pump
   d) None of the above

15) This invention:

a) Has no need for a cartridge seal
b) Is of simple design
c) Is axially compact
d) All of the above

16) The "Background of the Invention" describes:

a) The purpose of vehicle coolant pumps
b) The operation of vehicle coolant pumps
c) The problems with the cartridge seal
d) All of the above

17) U.S. Patent 3,632,220 is cited because it:

a) Describes problems with the cartridge seal
b) Has a magnetic drive
c) Is axially compact
d) All of the above

18) U.S. Patent 3,802,804 is cited because it:

a) Does not have a magnetic drive
b) Is not powered by an electric motor
c) Does not have strict space limitations
d) None of the above

19) U.S. Patent 4,304,532 is cited because it:

a) Does not have a magnetic drive
b) Is motor driven
c) Has strict space limitations
d) All of the above

20) U.S. Patent 4,115,040 is cited because:

a) It has a magnetic drive
b) It has no space limitations
c) Is motor driven
d) All of the above

21) The "Summary of the Invention":

a) Explains every facet of every part of the invention
b) So there will be no mistake about the invention
c) Or no mistake of patentable features
d) All of the above

22) The "Description of the Embodiment":

a) Claims the pulley and hub as separate pieces
b) Claims the pulley and hub as one piece
c) Claims the double row "rolling" bearing
d) All of the above

# ANSWER KEY

1) D
2) D
3) C
4) A
5) D
6) B
7) C
8) D
9) C
10) A
11) C
12) D
13) D
14) B
15) D
16) D
17) A
18) C
19) B
20) D
21) D
22) D

# Learning Objectives

This course teaches the following specific knowledge and skills:

- The definition, purpose, and benefits of a patent
- The history of patents both abroad and in the United States
- The qualifications needed for a patent
- Limitations of a patent because of the rights of other adjoining patent holders
- The nature of patent infringement court cases
- The various dispositions that can be made of a patent
- An explanation of the main features of the new United States patent law signed into effect September 16, 2011.
- The three different types of patents that are granted
- The step-by-step procedure in preparing a patent application and working with a Patent Attorney
- Examination of a patent granted to a major U.S. car company
- The operation of an auto engine cooling system
- The importance of the "cartridge seal" in an automobile engine coolant pump
- The purpose of the "Abstract" in a patent document
- The purpose of the "Background of the Invention" part of a patent document
- The reason for citing other similar patents (Prior Art) in a patent document and how they are used as an advantage
- The explanation of the "Description of the Preferred Embodiment" part of a patent document
- How patent claims can be used to protect a patent from infringement

# Overview

This course begins describing patents and the patent application process. It then examines all aspects of a patent assigned to a major U.S. automobile manufacturer to improve the performance of one of its products. It explains how four other patents examined as "Prior Art" were used to enhance the credibility of the sought after patent and how the three claims that were made skillfully protect against possible infringement.

This course contains information that every engineer should be made aware of. It teaches how to protect innovative ideas, how to apply for a patent, and how to use "Prior Art" and claims in obtaining a patent and protecting it from possible infringement. This course is intended for all fields of engineering and any other parties of interest.

# About the Author

The author has a BSME from Case-Western University, Cleveland, Ohio. He is a registered Professional Engineer in the State of Ohio. He has had 40 years of Mechanical Engineering experience, 26 of which were with the General Motors Corporation. While there, he obtained U.S. Patent number 4,645,432, "Magnetic Drive Vehicle Coolant Pump". He went on to become a leader in anti-friction bearing applications in both the automotive and industrial fields. Valuable experience was also gained in gears and mechanical power transmission. Prior to that he was employed by TRW, Cleveland, Ohio, where he was responsible for bearings, gears and mechanical power transmission in the aircraft and missile fields under the tutelage of Mr. Thomas Barish, a leading mechanical power transmission consultant. Also, Mr. Tata has authored 25 technical papers that are available on the internet and other sources for professional development hours. He is also the author of the book "The Development of U.S. Missiles During the Space Race with the U.S.S.R.". It is based on his experience, early in his career, working as a Flight Test Engineer at Cape Canaveral, Florida during the Cold War with the U.S.S.R. More recently, Mr. Tata has ventured outside the technical field in authoring his second book, "The Greatest American Presidents". Following that is his third work, a part technical, part historical book titled "How Detroit became the "Automotive Capital of the World". The fourth book is a workbook sized publication titled "Mechanical Engineering Primer" complete with a multiple choice quiz for classroom use or any other party so inclined.

# PART 2

# CASE STUDY II

# CONTENTS
## PART 2

# INTRODUCTION

Part 1 of this course examined patents, the patent application process, and a patent granted to a U.S. corporation to improve one of its products. This course, Part 2, Case Study II, examines a patent application that reveals another side of the patent process.

A patent, as mentioned, is a document issued by the government that gives the engineering applicant exclusive rights to an invention for a period of 20 years. The invention must be something that is new and useful to society. An invention that is obvious to a person with average skill in a given field is not patentable. Patent laws give individuals and corporations incentive to develop their innovative ideas for their own benefit and for the benefit of others.

There are three types of patents; utility patents, design patents, and plant patents. Engineering patents usually fall into the category of utility patents. Utility patents involve materials, machines and components, and manufacturing parts and processes. Design patents involve only the appearance of an article. Plant patents, as the name suggests, are given to those that reproduce a new variety of plant.

# Background

There has been a long standing debate among anti-friction bearing engineers as to what type of product to use in an application. This course deals with one of the arguments by examining an engineering patent request that suggests the use of angular contact ball bearings to replace tapered roller bearings in a mechanical power transmission application.

Angular contact ball bearings are similar to commonly used radial ball bearings except for the alignment of the balls and rings. Radial ball bearings have the balls and rings aligned to support radial loads. Radial loads act perpendicular to the bearing axis of rotation.

Angular contact ball bearings have the balls and rings aligned so that the line of contact is at an angle to the perpendicular. They are used to support both radial and thrust loads. Thrust loads act parallel to the bearing axis of rotation. (See figure 1, page 41)

Tapered roller bearings have the roller to ring contact line act at an angle to the perpendicular similar to angular contact ball bearings. Tapered roller bearings, like angular contact ball bearings, support both radial and thrust loads. (See figure 2, page 42)

Angular contact ball bearings operate with virtually pure rolling between the balls and rings. Tapered roller bearings, under load, create forces that act to "squirt" the roller large end against the inner ring (cone) flange face causing frictional losses that don't exist in angular contact ball bearings. (See figure 3, page 43)

Ball bearings have a much lower spring rate (load vs. deflection) than tapered roller bearings. The contact between balls and rings starts out as a point and, as the load increases, evolves into an ellipse. The contact for tapered roller bearings starts out as a line and, under increasing load, changes into a larger rectangular shape causing the spring rate of tapered roller bearings to be "orders of magnitude" greater than that of ball bearings.

The following pages describe the above mentioned patent application that was submitted for a gearshaft bearing arrangement. Included is a discussion of four existing patents that were selected as "prior art". Prior art are existing patents that are similar to the one being applied for. They can be used by the patent applicant as reasons why a patent should be allowed or they can used by the Patent Office as reasons why the patent should be denied.

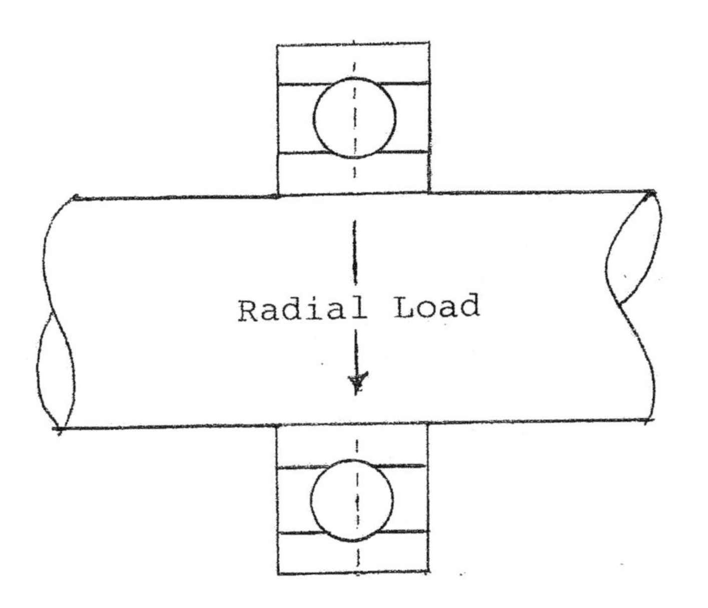


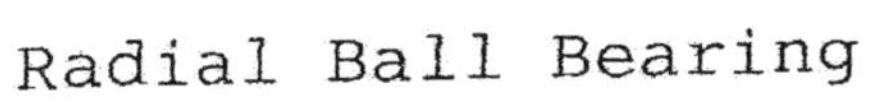


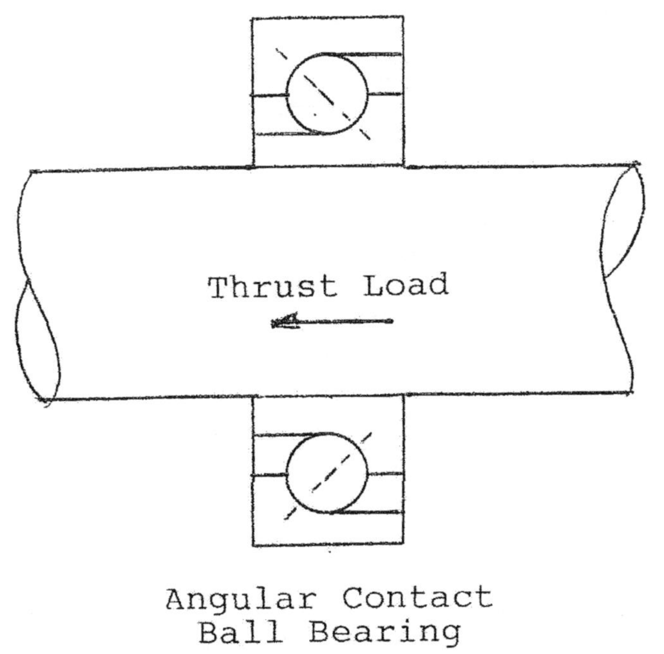


Angular Contact
Ball Bearing

**Figure 2**

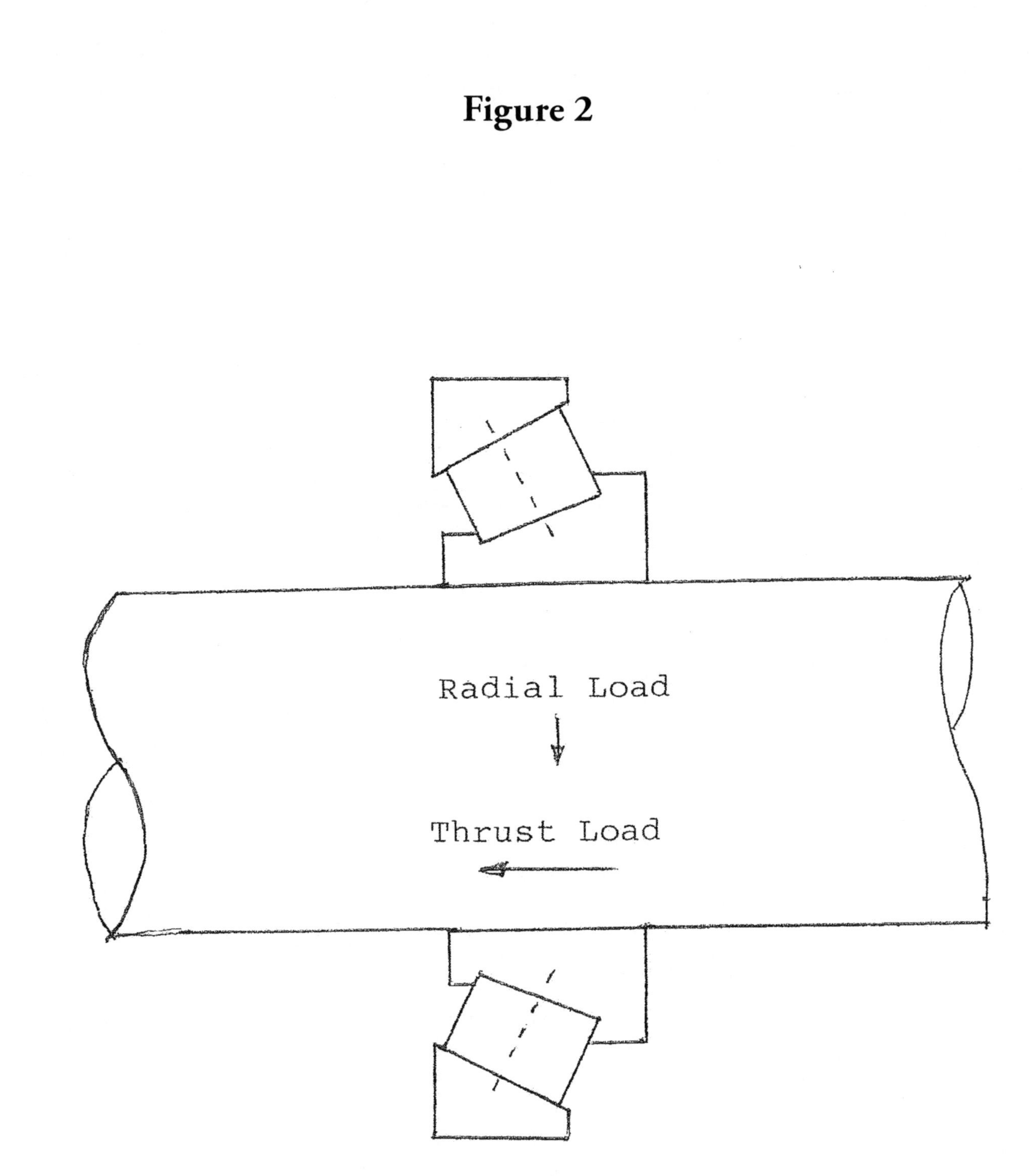


Tapered Roller Bearing

## Figure 3

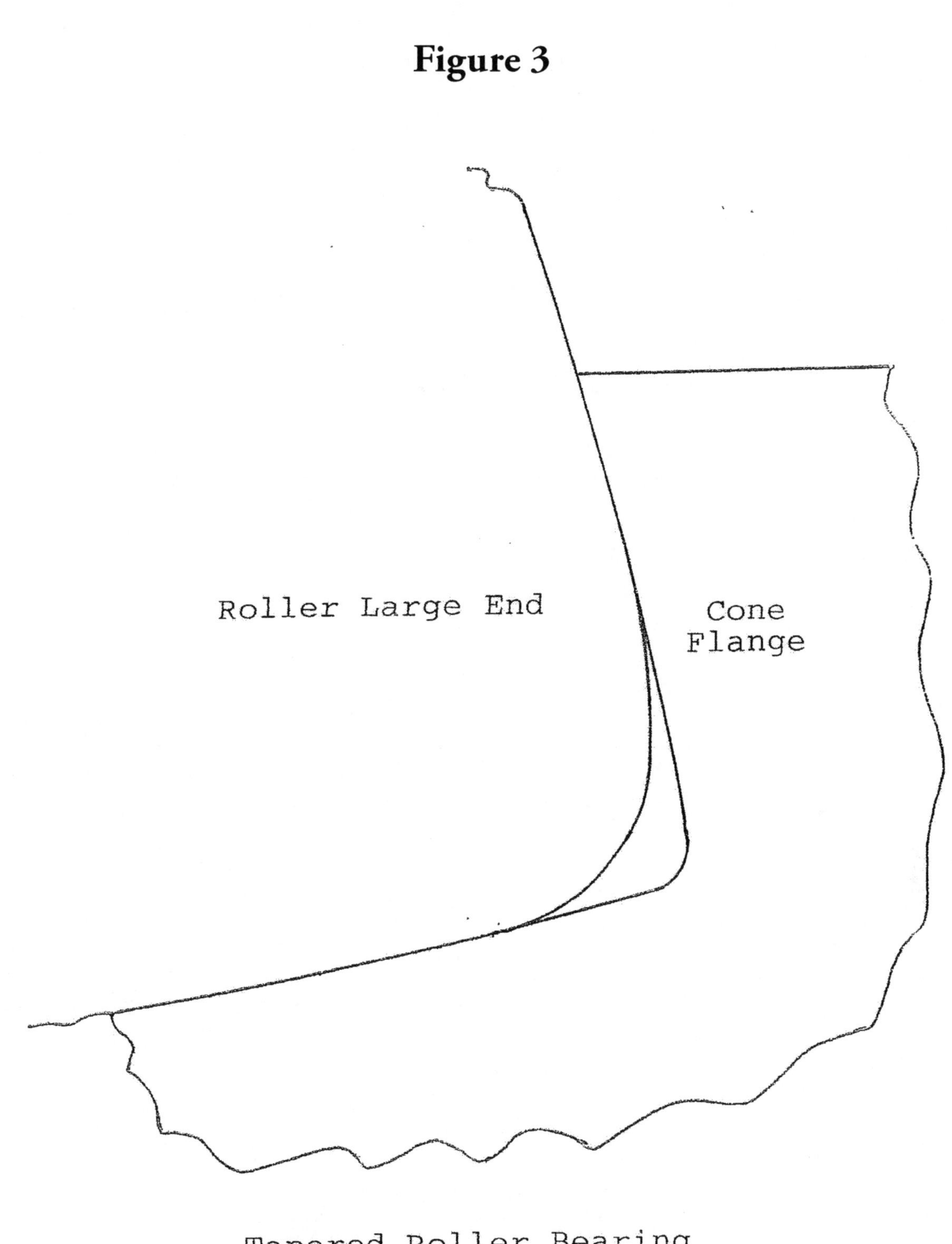


Tapered Roller Bearing
Roller End Contact

# Tapered Roller Bearing Gearshaft

Figure 4 (page 47) has a cross-sectional view of a conventional gearshaft arrangement. The assembly consists of output gear 1, tapered roller bearings 2, housing 3, gearshaft 4, bearing preload locknut 5, and collapsible spacer 6. The unit is lubricated with gearbox lube which is contained in the assembly by seal 7 and O-ring 8. Flange 9 connects to the driving member. If the output shaft is exposed to the elements, slinger 10 is used to limit foreign element intrusion into the seal area.

Although this type of bearing arrangement has proven successful in some applications, it is not without its disadvantages. Gearshaft support bearings have to be preloaded in order to support the shaft with enough rigidity to keep the gears in the proper mesh under all load conditions. To preload gearshaft bearings, an axial load (thrust) is imposed on one bearing that is reacted by the other. On Figure 4, by torqueing locknut 5, a thrust load is imposed on the right bearing inner ring (cone). At the same time, the same thrust load is transferred through the housing to the left bearing outer ring (cup) putting the two bearings into compressive preload. As previously discussed, ball bearings have a low spring rate because of having a smaller contact pattern between the balls and rings while tapered roller bearings have a high spring rate because of having a much larger contact pattern between the rollers and rings. This makes it easy to apply and control preload on ball bearings and very hard to apply and control preload on tapered roller bearings. Too much preload results in potential bearing failure and too little preload results in potential gear failure. The problem is so acute with tapered roller bearings that they are not normally recommended to be run under preload conditions. Since tapered roller bearings are used in this application and preloading is required, special steps have to be taken for the design to be successful.

During assembly, a unique procedure is used on a special machine in order to set the preload on bearings 2 to the amount needed to support the shaft with enough rigidity to insure proper gear mesh under all power conditions. On this machine, while the gear shaft is rotated, adjusting nut 5 is rotated slightly faster in the same direction until a predetermined torque level is reached. This torque level corresponds to the correct amount of preload that is required for the application. Because tapered bearings are so stiff to the application of load, collapsible spacer 6 must be used to control the rise in torque while the nut is being rotated in order to accurately set the preload.

There is another factor against the use of tapered bearings in this application. Ball bearings operate with virtual pure rolling; however, when tapered roller bearings are under load, the roller large end is "squirted" or forced against the cone (inner ring) flange face causing sliding friction. (This roller inner ring flange interface is one of the most important elements in the design of tapered roller bearings.) This contact produces additional bearing torque and is subject to wear from gear debris and poor lubrication which, in time, cause the loss of preload with the resulting adverse effect on the gear mesh. Special precautions in the design and manufacture of the tapered roller large end and the cone flange face have to be taken for the bearing to operate satisfactorily. Even then, the lubricant must be free of contamination from gear and other hard particle debris. Magnets are put in lubricant sumps in an effort to attract ferrous particles. The lubricant has to be changed after a period of time because of thermal breakdown.

In conclusion, it can be seen that, to have successful operation, extra care has to be taken when using shafts designed with tapered roller bearings.

The following pages will examine a design that a patent was applied for using ball bearings to support the gearshaft. Ball bearings, as explained above, are more easily preloaded and operate with less friction than tapered roller bearings. Included in the study will be an analysis of four patents that were investigated as "prior art" (patents that are similar to the one being applied for) that employ the use of ball bearings to support shafts.

## Figure 4

Tapered Roller Bearing Gearshaft

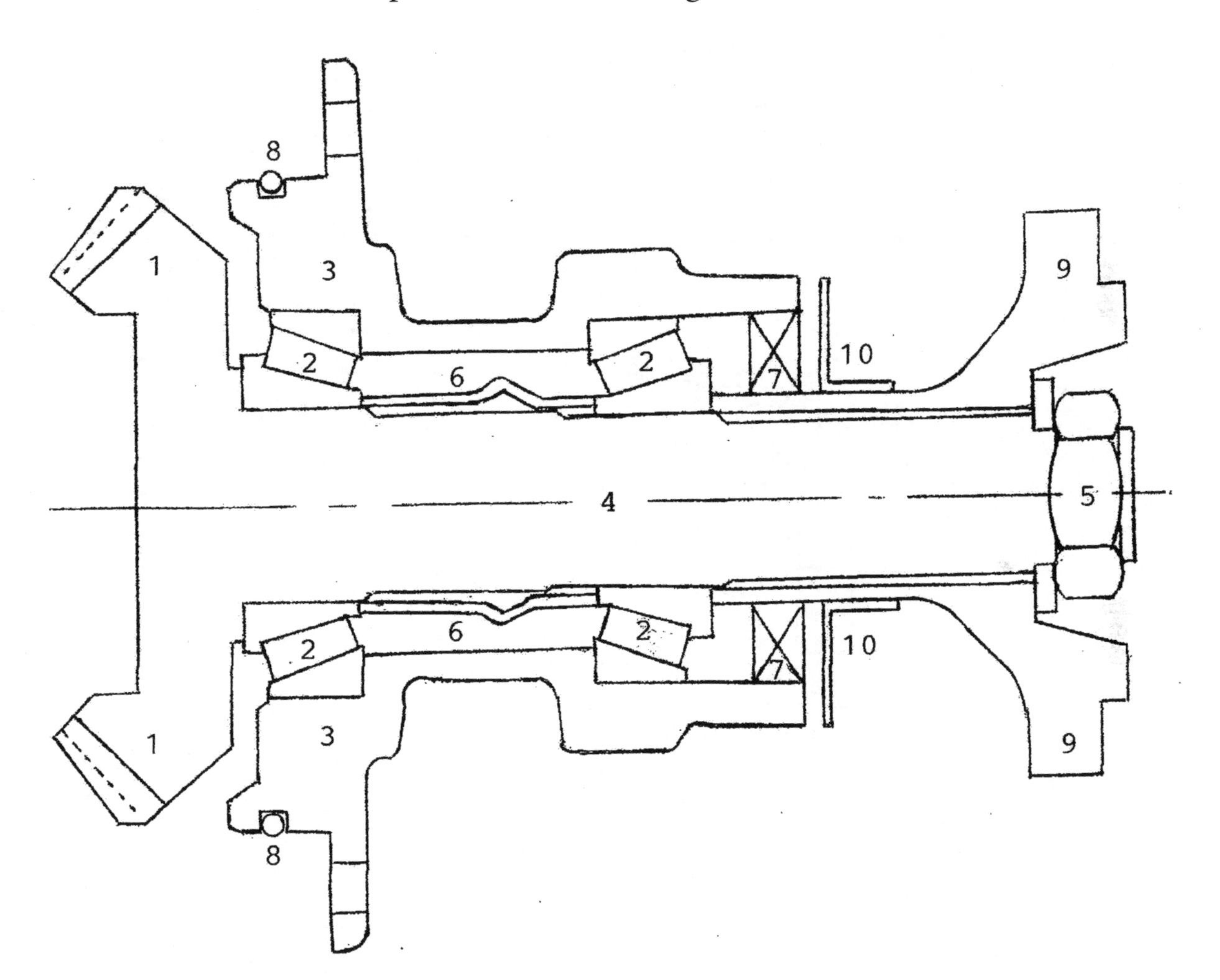

# Figure 5

Ball Bearing Gearshaft

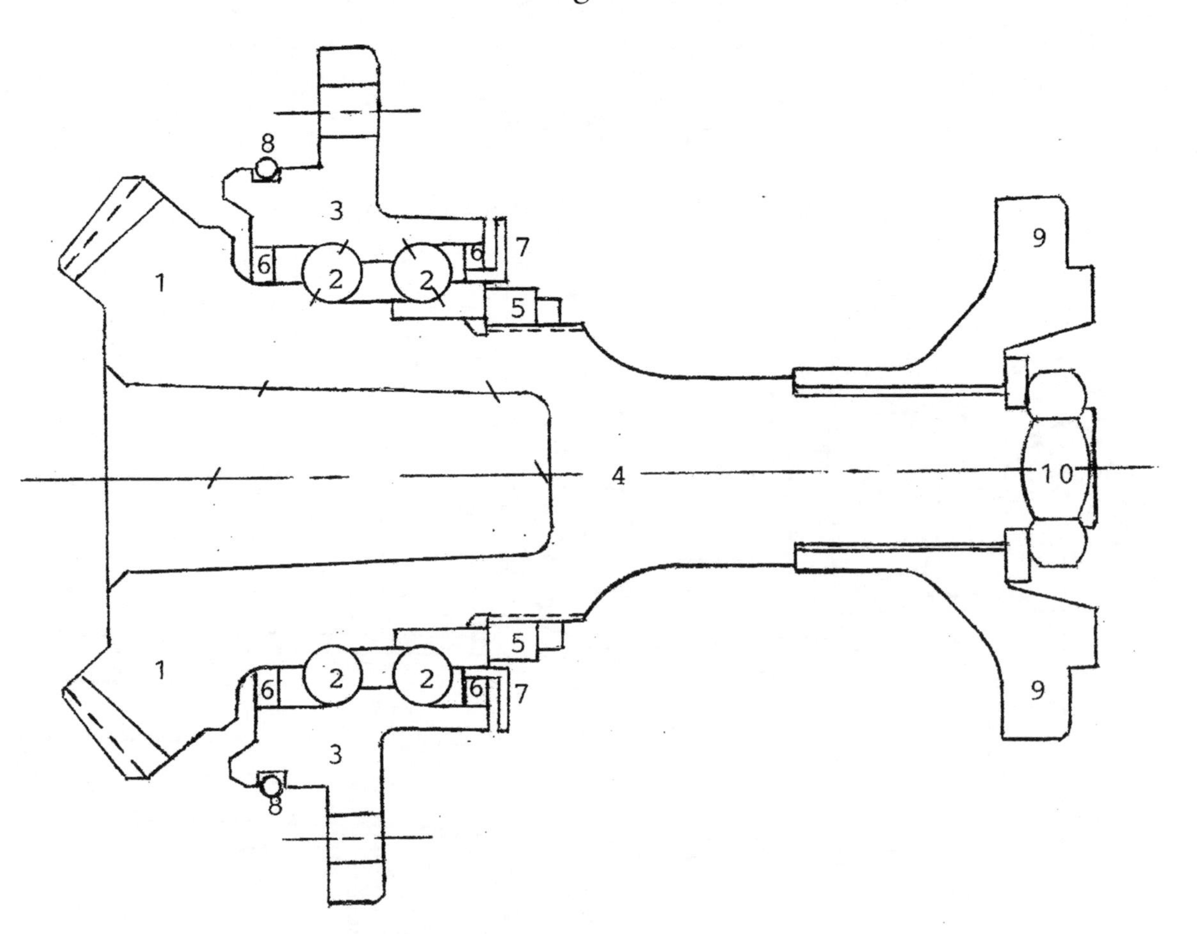

# Ball Bearing Gearshaft

Figure 5 above has a sketch of a gearshaft similar to Figure 4 except that the tapered roller bearings have been replaced with angular contact ball bearings. Item 1 is the gear. Item 2 are the two angular contact ball bearings. As shown on figure 2, angular contact ball bearings (as well as tapered roller bearings) have the line of contact between the balls and the two rings of each bearing lie at an angle extending outward from the vertical which aids in supporting overhanging loads such as the gear in this application. Item 3 is the hub. Item 4 is the gearshaft. Item 5 is the ball bearing clamping nut. Item 6 are the two bearing seals and item 7 is the slinger seal. Item 8 is an o-ring needed to prevent leakage of gearbox fluid. Item 9 is the rear drive shaft mounting flange and item 10 is the mounting flange retaining nut. Following are explanations of the claims that were given in the patent application describing the advantages of using angular contact ball bearings instead of tapered roller bearings to support gearshafts:

- Ball bearings have a very low spring rate compared to tapered roller bearings and therefore preload is more easily set and maintained under all gear load conditions. The preload for the two ball bearings in this design is ground in and automatically set by simply tightening locknut 5. This is a much more accurate and easier task than requiring the use of a special machine and a collapsible spacer to set the preload as is required for tapered roller bearings.
- Ball bearings operate with less friction than tapered roller bearings and with less chance of losing preload. In automotive applications, this decreases vehicle drive line friction increasing fuel miles per gallon. Ball bearings rotate with nearly perfect rolling as compared to tapered roller

bearings which have sliding friction occurring at the roller/inner ring interface. Also, gear debris and lubricant breakdown cause excessive wear at this interface resulting in loss of preload.

- The angular contact ball bearing design has three of the four ball pathway diameters ground directly in the hub and on gearshaft saving significant machining and assembly time when compared to the tapered roller bearing design where all inner and outer rings are separately manufactured and assembled components. One ball bearing inner ring is separable on each bearing so that a full complement of oversized balls can be assembled into each ball row greatly increasing the capacity of the design and its ability to support heavy gear loads.
- The angular contact ball bearings are designed with a larger diameter and with less separation between the two rows than the tapered roller bearings offering design compactness and more support for the gear because of a stronger housing and shaft.
- The ball bearing design is lubricated with grease and sealed against the entrance of gearbox lubricant which can become contaminated with hard particle debris. Tapered roller bearings do not operate as efficiently with grease lubrication as do ball bearings because of the inability of the lubricant to satisfactorily penetrate the roller/inner ring interface.

# PRIOR ART

Listed below are copies of the first page of four inventions that were investigated as prior art for the subject patent application:

1) Figure 6, U.S. Patent 4,248,487: This patent shows a compromised version of the means of supporting a gearshaft by using one tapered roller bearing row and one angular contact ball bearing row. This combination of bearings supporting a gearshaft is not commonly used and is shown for informational purposes only.

2) Figure 7, U.S. Patent 3,594,051: Although this patent is concerned with the method of retaining angular contact ball bearings with a special spindle nut (36), it does show that supporting a spindle (shaft) with angular contact ball bearings is not new; however, the spindle supports an overhung wheel, not an overhung gear, which makes the application of this patent unlike the one being requested. The patent being requested may still be allowed based on the difference in applications; however, it does present a weaker argument than if the idea were to be novel to all applications.

3) Figure 8, U.S. Patent 3,792,625: This patent is concerned with the method used in preloading two angular contact ball bearings. The bearings support an overhung gearshaft similar to the one in the subject application which greatly weakens the main claim of the patent being sought after. The method of setting the preload differs in that a push on flanged member is used instead of a threaded spindle nut.

4) Figure 9, U.S. Patent 2,174,262: This patent dates back to 1939 and again has angular contact ball bearings (54&55) supporting a gearshaft. Ball pathways are ground directly on adjoining parts (49&50) and bearings are retained with a spindle nut (56) similar to the method used in the subject patent application.

The three prior art inventions show that supporting gearshafts with angular contact ball bearings having ground-on pathways and nut preloads have been in existence for many years and is not patentable. The other claims made for the ball bearing design such as design compactness and grease lubrication are considered to be good accompanying features but, in themselves, are not novel and therefore not patentable items. Based on all the above, the subject patent application was denied.

# Figure 6

# United States Patent [19]

**Åsberg**

[11] **4,248,487**

[45] **Feb. 3, 1981**

[54] **ROLLING BEARING**

[75] Inventor: **Sture L. Åsberg**, Partille, Sweden

[73] Assignee: **SKF Nova AB**, Gothenburg, Sweden

[21] Appl. No.: **15,191**

[22] Filed: **Feb. 26, 1979**

**Related U.S. Application Data**

[63] Continuation of Ser. No. 850,909, Nov. 14, 1977, abandoned.

[30] **Foreign Application Priority Data**

Dec. 1, 1976 [SE] Sweden ............ 7613436
May 17, 1977 [SE] Sweden ............ 7705758

[51] **Int. Cl.**[3] ............ **F16C 19/49**; F16C 23/06

[52] **U.S. Cl.** ............ **308/189 R**; 308/207 R; 308/236

[58] **Field of Search** ............ 308/177, 178, 183, 189 R, 308/189 A, 196, 207 R, 207 A, 211, 214, 236

[56] **References Cited**

U.S. PATENT DOCUMENTS


| | | | |
|---|---|---|---|
| 1,099,571 | 6/1914 | Rennerfelt | 308/211 |
| 1,476,329 | 12/1923 | Duesenberg | 308/189 A |
| 1,503,849 | 8/1924 | Proctor | 308/207 A |
| 1,511,905 | 10/1924 | Oldham | 308/211 |
| 2,010,965 | 8/1935 | Scrivener | 308/189 R |
| 2,037,982 | 4/1936 | Hughes | 308/236 |
| 2,438,542 | 3/1948 | Cushman | 308/196 |
| 2,551,503 | 5/1951 | Needham | 308/236 |
| 2,692,805 | 10/1954 | Maxwell | 308/189 R |
| 4,150,468 | 4/1979 | Harbottle | 308/236 |


*Primary Examiner*—Richard R. Stearns
*Attorney, Agent, or Firm*—Eugene E. Renz, Jr.

[57] **ABSTRACT**


The combination comprising a housing, a shaft journal having a gear wheel at one end, and a bearing assembly rotatably supporting the shaft journal in a cylindrical seat in the housing. The assembly includes a one-piece outer ring having threads on its outer periphery cooperating with threads in the cylindrical seat. The outer ring is rotatable to permit axial adjusting movement of the bearing assembly relative to the housing. The bearing includes two rows of rolling bodies in the annular space between the rings spaced closely relative to one another and an inner race ring on the shaft journal for each of the rows of rolling bodies. One of the rows comprises rollers having axes inclined at an angle to the bearing axis disposed adjacent the gear wheel and the other row of rolling bodies comprises balls which roll against raceways in the ring to provide angular contact disposed at the opposite end of the shaft journal. A locking member engages the threads of the outer ring to permit axial adjustment thereof relative to the housing and abutts the housing to lock the outer ring in a predetermined axial position in the housing.


**1 Claim, 2 Drawing Figures**

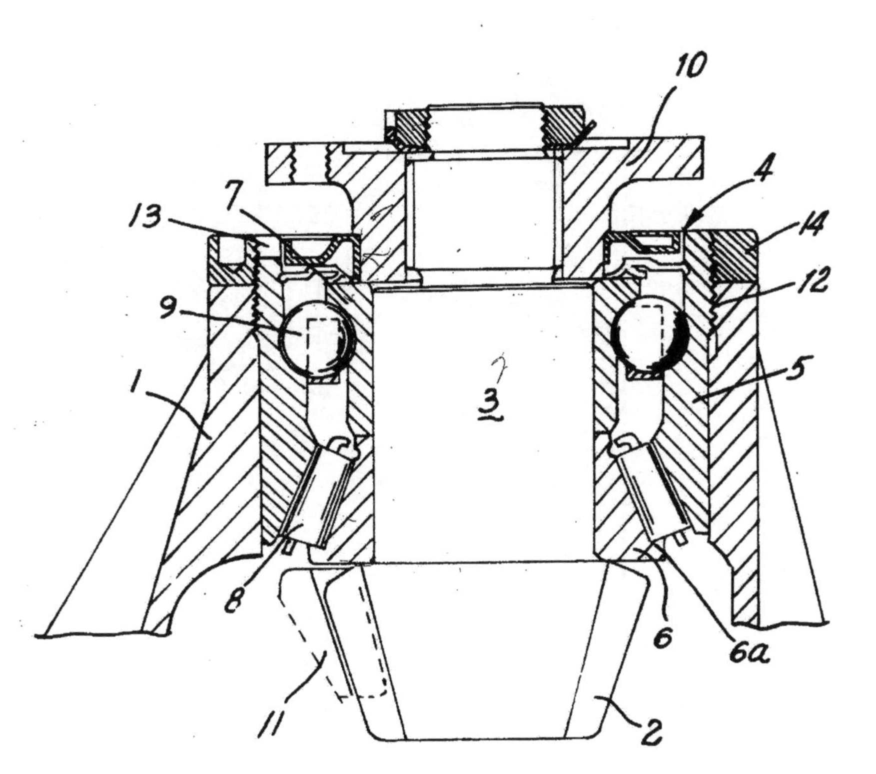

# Figure 7

## United States Patent

[11] **3,594,051**

[72] Inventor **Leonard A. Wells**
**Oklahoma City, Okla.**
[21] Appl No **844,107**
[22] Filed **July 23, 1969**
[45] Patented **July 20, 1971**
[73] Assignee **June M. Hicks**
**Oklahoma City, Okla.**

[54] **WHEEL BEARING MOUNTING**
**3 Claims, 2 Drawing Figs.**

[52] **U.S. Cl.** ........................................ **308/191**
[51] **Int. Cl.** ........................................ **F16c 13/02**
[50] **Field of Search** ........................................ 308/191, 207.1, 189.1, 189, 210, 211

[56] **References Cited**

UNITED STATES PATENTS

| | | | |
|---|---|---|---|
| 1,245,094 | 10/1917 | Douthit | 308/210 |
| 1,754,892 | 4/1930 | Hughes | 308/191 |
| 2,618,521 | 11/1952 | Shields | 308/211 |
| 2,622,934 | 12/1952 | Phelps | 308/210 |

*Primary Examiner*—Martin P. Schwadron
*Assistant Examiner*—Frank Susko
*Attorney*—Dunlap, Laney, Hessin & Dougherty

**ABSTRACT:** The present invention relates to an improved wheel bearing mounting of the type wherein one or more axially preloaded antifriction wheel bearings supporting a wheel hub are axially secured on a spindle by a spindle nut. The wheel bearing mounting of the present invention includes a spindle nut having a pair of aligned openings in opposite sides thereof for mating with a cotterway disposed in the threaded portion of the spindle. The spindle nut is of a size such that when the nut is threaded on the spindle to the position where the pair of aligned openings therein mate with the cotterway in the spindle, the desired axial preload on said bearings is provided.

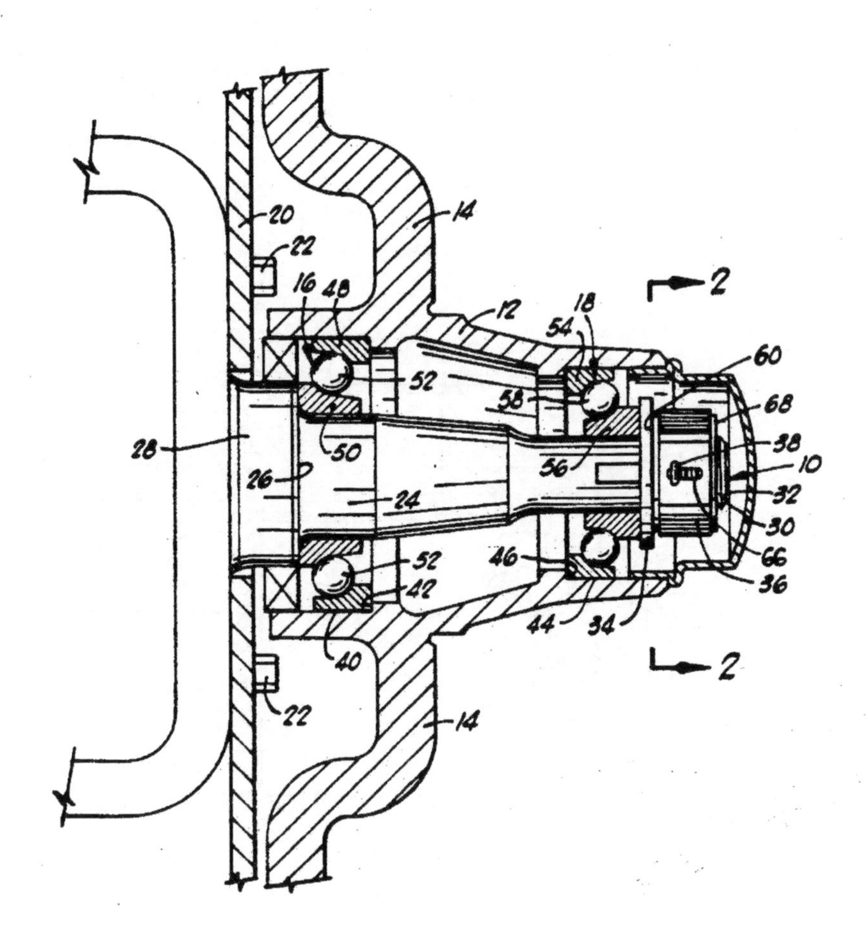

# Figure 8

**United States Patent** [19]
**Åsberg**

[11] **3,792,625**
[45] **Feb. 19, 1974**

[54] **PINION GEAR TRANSMISSION**

[75] Inventor: **Sture Åsberg,** Savedalen, Sweden

[73] Assignee: **SKF Industrial Trading and Development Co. N.V.,** Overtoom 141-145, Amsterdam, Netherlands

[22] Filed: **June 28, 1971**

[21] Appl. No.: **157,594**

[52] **U.S. Cl.** .................................................. **74/424**
[51] **Int. Cl.** ............................................... **F16h 1/14**
[58] **Field of Search** ............. 74/423, 424, 710, 713; 308/187, 191, 193, 195

[56] **References Cited**

UNITED STATES PATENTS

| | | | |
|---|---|---|---|
| 1,956,237 | 4/1934 | Hughes | 74/424 |
| 3,375,727 | 4/1968 | Nasvytis et al. | 74/423 X |
| 783,168 | 2/1905 | Baker | 74/713 |
| 899,891 | 9/1908 | Niclausse | 74/713 |
| 3,385,133 | 5/1968 | Terao | 74/710 |
| 3,290,101 | 12/1966 | Recknagel | 308/187 |
| 1,835,525 | 12/1931 | Robbins | 74/710 X |
| 1,369,210 | 2/1921 | Zimmerman | 74/424 X |
| 1,708,710 | 4/1929 | Vincent | 74/424 |

*Primary Examiner*—Benjamin W. Wyche
*Assistant Examiner*—Frank H. McKenzie, Jr.
*Attorney, Agent, or Firm*—Howson & Howson; Eugene E. Renz, Jr.

[57] **ABSTRACT**

Pinion gear transmission comprising a pinion shaft, a pinion wheel at one end of said shaft, connecting means at the opposite end of said shaft for connecting said pinion wheel and shaft to drive means, a housing, bearing means for rotatably supporting said pinion shaft and connecting means relative to said housing, said pinion shaft and connecting means movable axially relative to one another to preload said bearing means and means for permanently fixing said shaft and connecting means with said bearing means preloaded, said shaft, bearing means and connecting means forming an integral unit.

**17 Claims, 5 Drawing Figures**

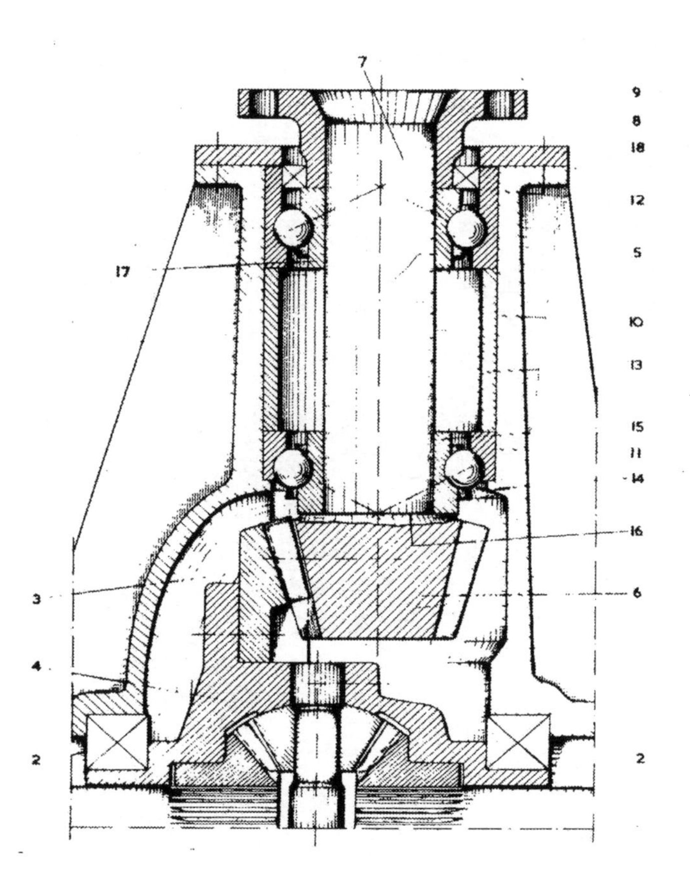

# Figure 9

**Sept. 26, 1939.** W. R. GRISWOLD **2,174,262**

METHOD OF MAKING BEARINGS

Original Filed March 20, 1935 2 Sheets—Sheet 1

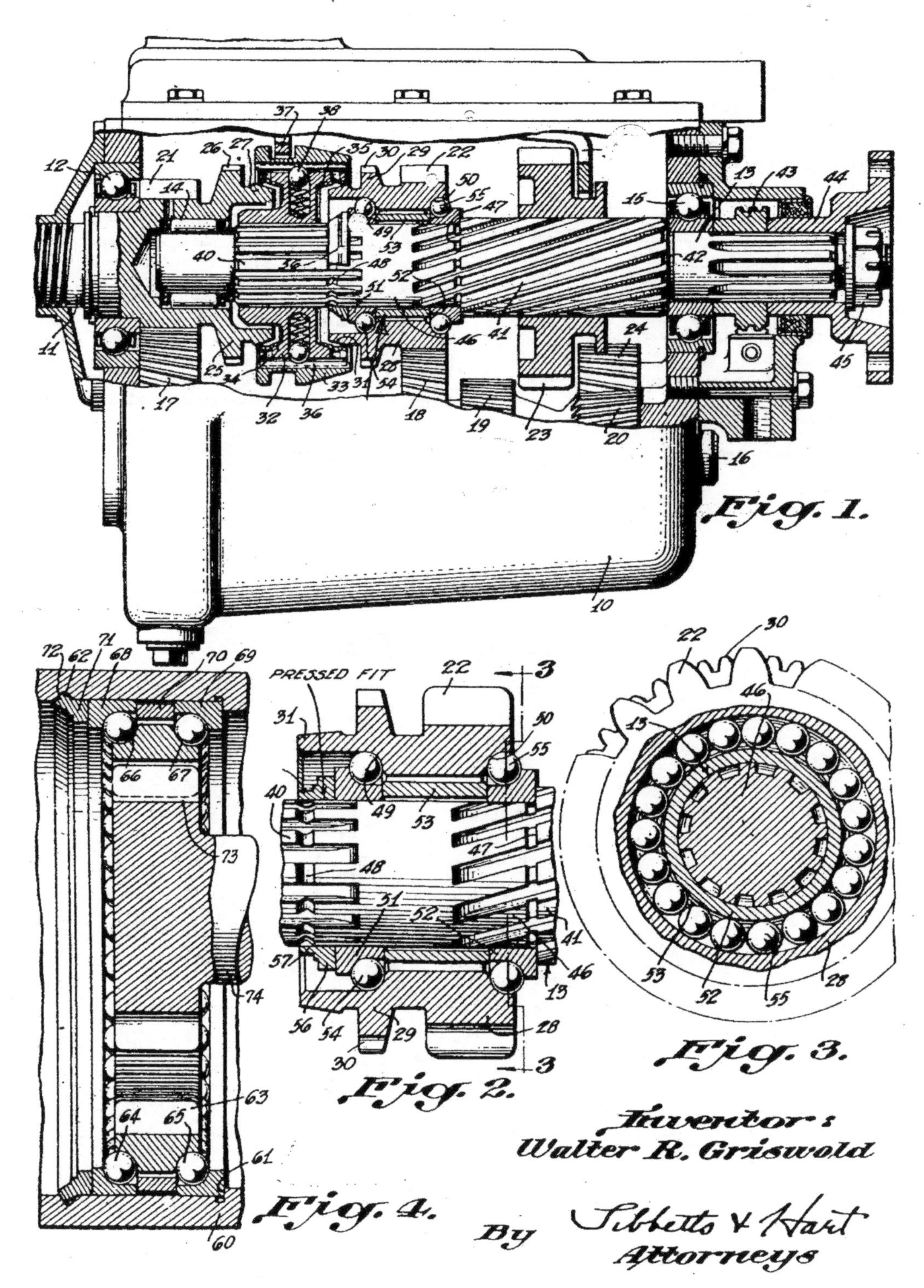

# QUIZ

1) A patent must be:
    a) New
    b) Not obvious
    c) Useful to society
    d) All of the above

2) The three types of patents are:
    a) Utility
    b) Design
    c) Plant
    d) All of the above

3) Utility patents involve:
    a) Materials
    b) Machines
    c) Processes
    d) All of the above

4) Angular contact ball bearings support:
    a) Radial loads only
    b) Thrust loads only
    c) Radial and thrust loads
    d) Axial loads only

5) Tapered roller bearings support:
    a) Radial and thrust loads
    b) Radial loads only
    c) Thrust loads only
    d) Axial loads only

6) Tapered bearing roller end contact
    a) Lowers bearing efficiency
    b) Raises bearing efficiency
    c) Does not affect bearing efficiency
    d) Exists only under heavy preload

7) Ball bearings operate at efficiencies that are:
    a) The same as tapered roller bearings
    b) Lower than tapered roller bearings
    c) Higher than tapered roller bearings
    d) In the high 80% efficiency range

8) Ball bearing spring rate is:
    a) Higher than tapered roller bearings
    b) Lower than tapered roller bearings
    c) The same as tapered roller bearings
    d) None of the above

9) Ball bearing load contact pattern forms:
    a) A square
    b) A rectangle
    c) An ellipse
    d) A triangle

10) Tapered roller bearing load pattern forms:

   a) A square
   b) A rectangle
   c) An ellipse
   d) A triangle

11) Prior art is:

   a) Patents obtained by a famous Englishman
   b) Art created before a patent is allowed
   c) Similar patents obtained prior to one submitted
   d) None of the above

12) Figure 4 is a view of a gearshaft using:

   a) Ball bearings
   b) Roller bearings
   c) Needle bearings
   d) Tapered roller bearings

13) Tapered roller bearing gearshafts include:

   a) Bearings spread less than angular contacts
   b) Larger diameter bearings than angular contacts
   c) The use of a collapsible spacer
   d) All of the above

14) A collapsible spacer is used with:

   a) Ball bearings
   b) Tapered roller bearings
   c) Sleeve bearings
   d) Thrust ball bearings

15) Ball bearings are:
   a) Less easily preloaded than tapered rollers
   b) More easily preloaded than tapered rollers
   c) Always preloaded
   d) Never preloaded

16) Features of the ball bearing gearshaft include:
   a) No collapsible spacer
   b) Grease lubrication
   c) Pathways ground directly in hubs and on shafts
   d) All of the above

17) The shaft on figure 6 uses:
   a) Two tapered roller bearings
   b) Two angular contact ball bearings
   c) A tapered roller and an angular ball bearing
   d) A tapered roller and a cylindrical roller bearing

18) The shaft on figure 7 supports:
   a) A machine tool
   b) A boring tool
   c) A wind turbine
   d) None of the above

19) The shaft on figure 8:
   a) Is different than the patent being submitted
   b) Is very similar to the one being submitted
   c) Can be used as a claim.
   d) Is supported by tapered roller bearings

20) The shaft on figure 9:

a) Is different than the patent to be submitted
b) Is similar to the patent being submitted
c) Can be used as a claim
d) Is supported by tapered roller bearings

# ANSWER KEY

1) D
2) D
3) D
4) C
5) A
6) A
7) C
8) B
9) C
10) B
11) C
12) D
13) C
14) B
15) B
16) D
17) C
18) D
19) B
20) B

# Learning Objectives

- The definition of a patent
- The requirements of a patent
- The types of patents
- The nature of loads on anti-friction bearings
- The two kinds of ball bearings
- The features of tapered roller bearings
- The friction characteristics of ball versus tapered roller bearings
- The definition of "prior art"
- The design of a tapered roller bearing gearshaft
- The design of an angular contact ball bearing gearshaft
- The definition of "prior art" and how it is used in obtaining a patent

# Overview

This document, Part 2, Case Study II, teaches both patent application and the important design and operating characteristics of two different types of anti-friction bearings. The patent application involves replacing conventionally used tapered roller bearings with angular contact ball bearings in a mechanical setting. It details important design features of ball and roller bearings and how they affect their installation and operation. Prior art (patents that are similar to the one being sought after) is examined and a conclusion drawn as to whether the patent application is to be allowed or denied based on the designs shown on prior art. This course is designed for technical personnel of all disciplines including school age students. Besides gaining general anti-friction bearing knowledge, lessons can be learned on how to apply for patents and how to examine prior art for claims made in already granted patent applications.

# INSTRUCTOR BIOGRAPHY

The author has a BSME from Case-Western University, Cleveland, Ohio. He is a registered Professional Engineer in the State of Ohio. He has had 40 years of Mechanical Engineering experience, 26 of which were with the General Motors Corporation. While there, he obtained U.S. Patent number 4,645,432, "Magnetic Drive Vehicle Coolant Pump". He went on to become a leader in anti-friction bearing applications in both the automotive and industrial fields. Valuable experience was also gained in gears and mechanical power transmission. Prior to that he was employed by TRW, Cleveland, Ohio, where he was responsible for bearings, gears and mechanical power transmission in the aircraft and missile fields under the tutelage of Mr. Thomas Barish, a leading mechanical power transmission consultant. Also, Mr. Tata has authored 25 technical papers that are available on the internet and other sources for professional development hours. He is also the author of the book "The Development of U.S. Missiles During the Space Race with the U.S.S.R.". It is based on his experience, early in his career, working as a Flight Test Engineer at Cape Canaveral, Florida during the Cold War with the U.S.S.R. More recently, Mr. Tata has ventured outside the technical field in authoring his second book, "The Greatest American Presidents". Following that is his third work, a part technical, part historical book titled "How Detroit became the "Automotive Capital of the World". The fourth book is a workbook sized publication titled "Mechanical Engineering Primer" complete with a multiple choice quiz for classroom use or any other party so inclined.